Everyday Mathematics®

The University of Chicago School Mathematics Project

Student Math Journal
Volume 1

Grade 4

 Education

Chicago, IL • Columbus, OH • New York, NY

The University of Chicago School Mathematics Project (UCSMP)

Max Bell, Director, UCSMP Elementary Materials Component; Director, *Everyday Mathematics* First Edition; James McBride, Director, *Everyday Mathematics* Second Edition; Andy Isaacs, Director, *Everyday Mathematics* Third Edition; Amy Dillard, Associate Director, *Everyday Mathematics* Third Edition; Rachel Malpass McCall, Associate Director, *Everyday Mathematics* Common Core State Standards Edition

Authors
Max Bell, John Bretzlauf, Amy Dillard, Robert Hartfield, Andy Isaacs, Rebecca W. Maxcy†, James McBride, Kathleen Pitvorec, Peter Saecker, Robert Balfanz*, William Carroll*, Sheila Sconiers*

*First Edition only   †Common Core State Standards Edition only

Technical Art
Diana Barrie

Photo Credits
Front Cover (l)Tony Hamblin/Frank Lane Picture Agency/CORBIS, (r)Gregory Adams/Lonely Planet Images/Getty Images, (bkgd)John W Banagan/Iconica/Getty Images; **Back Cover Spine** Gregory Adams/Lonely Planet Images/Getty Images; **iii iv** The McGraw-Hill Companies; **v** Nick Rowe/Photodisc/Getty Images; **vi vii viii** The McGraw-Hill Companies; **22** Ryan McVay/Getty Images.

Third Edition Teachers in Residence
Rebecca W. Maxcy, Carla L. LaRochelle

UCSMP Editorial
Laurie K. Thrasher, Kathryn M. Rich

Contributors
Martha Ayala, Virginia J. Bates, Randee Blair, Donna R. Clay, Vanessa Day, Jean Faszholz, James Flanders, Patti Haney, Margaret Phillips Holm, Nancy Kay Hubert, Sybil Johnson, Judith Kiehm, Carla LaRochelle, Deborah Arron Leslie, Laura Ann Luczak, Mary O'Boyle, William D. Pattison, Beverly Pilchman, Denise Porter, Judith Ann Robb, Mary Seymour, Laura A. Sunseri

 This material is based upon work supported by the National Science Foundation under Grant No. ESI-9252984. Any opinions, findings, conclusions, or recommendations expressed in this material are those of the authors and do not necessarily reflect the views of the National Science Foundation.

everydaymath.com

STEM McGraw-Hill is committed to providing instructional materials in Science, Technology, Engineering, and Mathematics (STEM) that give all students a solid foundation, one that prepares them for college and careers in the 21st century.

Send all inquiries to:
McGraw-Hill Education
STEM Learning Solutions Center
P.O. Box 812960
Chicago, IL 60681

ISBN: 978-0-07-657636-4
MHID: 0-07-657636-1

Printed in the United States of America.

2 3 4 5 6 7 8 9 QMD 17 16 15 14 13 12 11

Contents

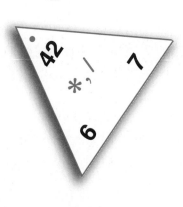
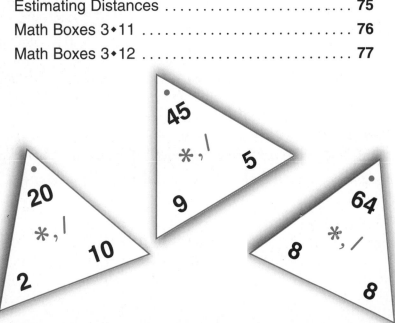

Contents **v**

UNIT 4 Decimals and Their Uses

UNIT 5 Big Numbers, Estimation, and Computation

UNIT 6 Division; Map Reference Frames; Measures of Angles

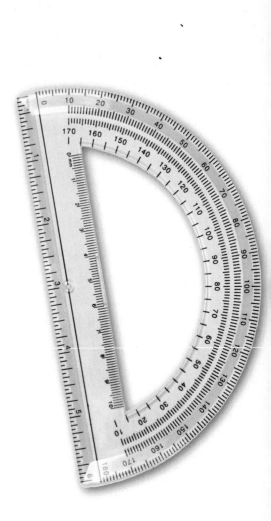

Projects

Activity Sheets

LESSON 1·1 Welcome to *Fourth Grade Everyday Mathematics*®

Much of your work in Kindergarten through third grade was basic training in mathematics and its uses. You learned to solve number stories and use arithmetic, including basic addition and multiplication facts.

Fourth Grade Everyday Mathematics builds on this basic training and begins to make the transition to mathematics concepts and ways of using mathematics that are more like what your parents and older siblings may have learned in high school. The authors believe that fourth graders can do more than was thought possible in years past.

Here are some things you will be asked to do in *Fourth Grade Everyday Mathematics:*

◆ Increase your "number sense," "measure sense," and estimation skills.

◆ Extend your skills in the basics of arithmetic—addition, subtraction, multiplication, and division. There is not much more to learn about the arithmetic of whole numbers, but over the next couple of years, you will become comfortable using fractions, percents, and decimals.

◆ Learn about variables (letters that stand for numbers) and other introductory topics in algebra.

◆ Develop your geometry concepts and skills with more exact definitions and classification of geometric figures, constructions and transformations of figures, and investigation of areas and volumes of shapes.

◆ Take a World Tour. Along the way, you will consider many kinds of data about various countries and learn how to use coordinate systems to locate places on world globes and maps.

◆ Do many projects involving numerical data.

In fourth grade, you will be asked to do more independent reading and investigation (often working with partners or in groups) rather than being told everything by your teacher.

We hope that you find the activities fun and that you see the beauty in mathematics. Most importantly, we hope you become better and better at using mathematics to solve interesting problems in your life.

LESSON 1·1

Using Your *Student Reference Book*

Use your *Student Reference Book* to complete the following:

1. Look up the word **mode** in the Glossary.

   a. Copy the definition. _____

   _____

   b. On which page in the *Student Reference Book* could you find more information

   about the mode of a set of data? page _____

2. Find the essay "Comparing Numbers and Amounts."

   a. Describe what you did to find the essay.

   _____

   _____

   b. Read the essay and solve the Check Your Understanding problems.

   Problem 1: _____   Problem 2: _____

   Problem 3: _____   Problem 4: _____

   c. Check your answers using the Answer Key.

3. Look up the rules for the game *Name That Number*.

   a. On which page did you find the rules? page _____

   b. How many players are needed for the game? _____ players

4. Go to the World Tour section. Record two interesting facts you find there.

   a. Fact 1: _____

   I found this information on page _____.

   b. Fact 2: _____

   I found this information on page _____.

LESSON 1·1

Math Boxes

---

1. Add mentally.

   a. 8 + 5 = _____

   b. 80 + 50 = _____

   c. 7 + 7 = _____

   d. 70 + 70 = _____

   e. 7 + 9 = _____

   f. 70 + 90 = _____

---

2. Fill in the missing numbers and state the rule.

   a. 2, 4, 6, __8__, _____, _____

   Rule: __+2__

   b. 65, 60, 55, _____, _____, _____

   Rule: _____

   c. 109, 95, 81, _____, _____, _____

   Rule: _____

160 161

---

3. Complete.

21 in. = _____ ft _____ in.

Circle the best answer.

A. 1 ft 1 in.

B. 1 ft 10 in.

C. 1 ft 9 in.

D. 1 ft 3 in.

129

---

4. Complete.

   a. 2 quarters = _____ dimes

   b. 1 dollar
and 5 nickels = _____ quarters

   c. 14 dimes = _____ pennies

   d. 8 quarters = _____ dollars

   e. 3 quarters
and 9 nickels = _____ dimes

---

5. Add mentally or with a
paper-and-pencil algorithm.

   a. 32 + 35 = _____

   b. 38 + 66 = _____

10 11

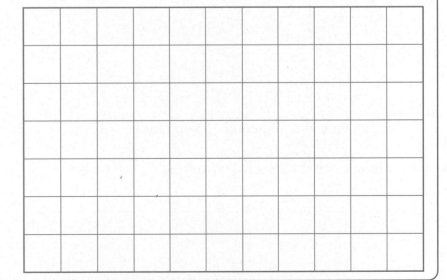

LESSON 1·2   Points, Line Segments, Lines, and Rays

Use a straightedge to draw the following:

1. a. Draw and label line segment RT ($\overline{RT}$).

    b. What is another name for $\overline{RT}$? _____

2. a. Draw and label line BN ($\overleftrightarrow{BN}$). Draw and label point T on it.

    b. What are 2 other names for $\overleftrightarrow{BN}$? _____

3. a. Draw and label ray SL ($\overrightarrow{SL}$). Draw and label point R on it.

    b. What is another name for $\overrightarrow{SL}$? _____

4. a. Draw a line segment from each point to each of the other points.

            M •   N •

        •   •
        O   P

    b. How many line segments did you draw? _____

    c. Write a name for each line segment you drew.

    _____

LESSON 1·2 — Math Boxes

1. Subtract mentally.

a. 9 − 4 = _____

b. 90 − 40 = _____

c. 16 − 9 = _____

d. 160 − 90 = _____

e. _____ = 17 − 8

f. _____ = 170 − 80

2. Draw and label line QR.
Draw point S on it.

What are two other names for line QR?

_____

_____

SRB
91

3. Complete.

Max read _____ books.

Sue read _____ books.

Ira read _____ books.

Pat read _____ books.

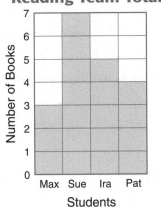

Reading Team Totals

SRB
76

4. Cross out the names that do not belong in the name-collection box. Label the box with the correct number.

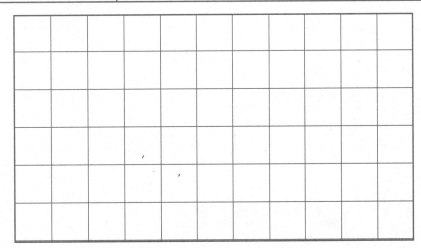

25 − 13
20 − 7
6 × 2
4 × 3
40 − 23
7 × 3

SRB
149

5. Subtract mentally or with a paper-and-pencil algorithm.

a. 86 − 21 = _____

b. 93 − 24 = _____

SRB
12–15

LESSON 1·3   Angles

1. Which angle is bigger,

∠*ABC* or ∠*DEF*? _____

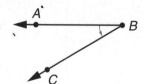

2. Draw ∠*BAC*. What is another name

for ∠*BAC*? _____

C•

3. What is the vertex of ∠*BAC*? Point _____

A•   •
              B

4. Feng said the name of this angle is ∠*SRT*. Is he right? Explain.

_____

_____

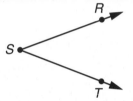

Try This

Use the points shown on the grid below and a straightedge to draw the angles.

5. Draw ∠*AED*.

   a. What is the vertex of the angle?   Point _____

   b. What is another name for ∠*AED*?   ∠ _____

6. Draw a right angle whose vertex is point *C*.   My angle is called ∠ _____.

7. Draw an angle that is smaller than a right angle.   My angle is called ∠ _____.

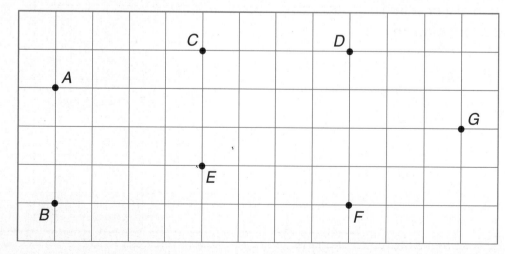

LESSON 1·3 **Addition and Subtraction**

Use your favorite addition and subtraction methods to solve the problems. Show your work.

1. 48 + 79 = _____

2. 392 + 75 = _____

3. _____ = 296 + 514

4. _____ = 937 + 652

5. _____ = 98 − 57

6. 122 − 82 = _____

7. _____ = 900 − 632

8. _____ = 512 − 239

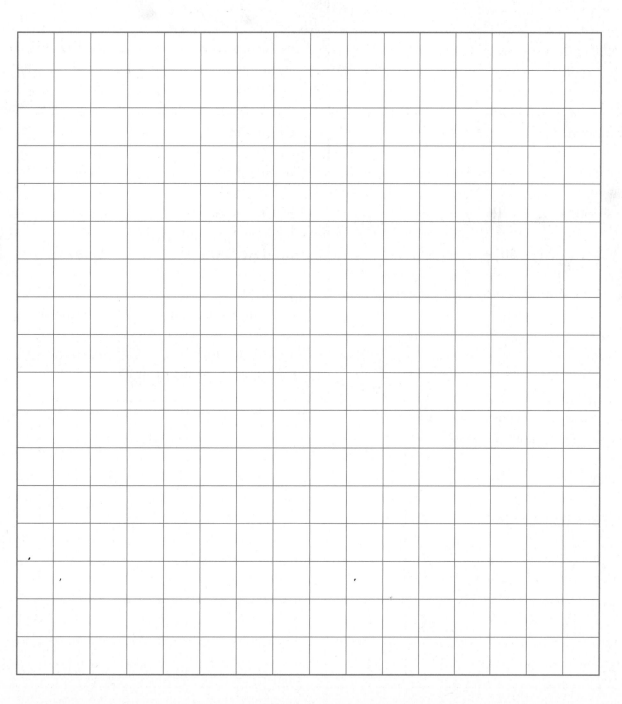

LESSON 1·3

Math Boxes

---

1. Add mentally.

   a. 6 + 7 = _____

   b. 60 + 70 = _____

   c. 9 + 8 = _____

   d. 90 + 80 = _____

   e. 4 + 7 = _____

   f. 40 + 70 = _____

2. Fill in the missing numbers and state the rule.

   a. 4, 8, 12, 16, _____, _____, _____

   Rule: _____

   b. 33, 30, 27, _____, _____, _____

   Rule: _____

   c. _____, _____, _____, 106, 141, 176

   Rule: _____

SRB 160 161

---

3. Complete.

   a. 1 ft = _____ in.

   b. 24 in. = _____ ft

   c. _____ yd = 36 in.

   d. 30 in. = _____ ft _____ in.

   e. 50 in. = _____ yd _____ ft

             _____ in.

SRB 129

4. Complete.

   a. 9 dimes = _____ pennies

   b. 30 dimes = _____ dollars

   c. 4 quarters = _____ dimes

   d. 2 dollars
       and 10 nickels = _____ quarters

   e. 13 dollars = _____ quarters

---

5. Add mentally or with a paper-and-pencil algorithm.

   a. 63 + 12 = _____

   b. 56 + 97 = _____

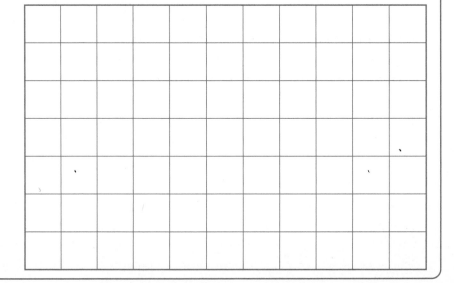

SRB 10 11

LESSON 1·4   Math Boxes

---

1. Subtract mentally.

a. $15 - 7 =$ _____

b. $150 - 70 =$ _____

c. $13 - 8 =$ _____

d. $130 - 80 =$ _____

e. $17 - 9 =$ _____

f. $170 - 90 =$ _____

---

2. Draw and label line *AB*.
Draw point *C* on it.

What are two other names for line *AB*?

_____

_____

---

3. Complete.

Luz sold _____ boxes.

Ana sold _____ boxes.

Mya sold _____ boxes.

Pei sold _____ boxes.

Cookie Sale

---

4. Which of these can go in a name-collection box for the number 50? Circle the best answer.

A. $10 + 35$

B. $136 - 51$

C. $200 \div 4$

D. 4×15

---

5. Subtract mentally or with a paper-and-pencil algorithm.

a. $76 - 41 =$ _____

b. $52 - 38 =$ _____

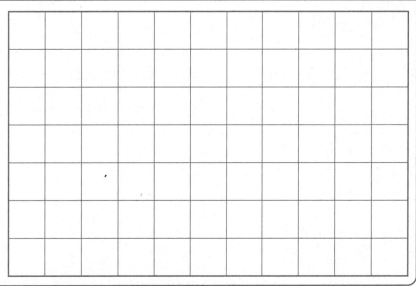

LESSON
1·4   # Parallelograms

1. Circle the pairs of line segments below that are parallel. Check some of your answers by extending each pair of segments to see if the two segments in the pair meet or cross.

a.

b.

c.

d.

e.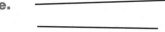

f.

Use your Geometry Template or straightedge to draw the following quadrangles:

2. Draw a quadrangle that has 2 pairs of parallel sides.

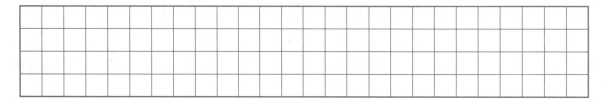

   This is called a _____.

3. Draw a quadrangle that has only 1 pair of parallel sides.

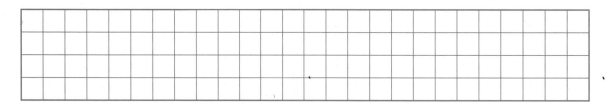

   This is called a _____.

LESSON 1·4   Parallelograms *continued*

For Problems 4 and 5, circle the best answer(s). Some items have more than
1 correct answer, so you may need to circle more than 1 answer.

4. A **parallelogram** is a quadrangle that
   has 2 pairs of parallel sides.
   Which are parallelograms?

   A. squares

   B. rectangles

   C. rhombuses

   D. trapezoids

5. A **rhombus** is a parallelogram in which
   all sides are the same length.
   Which are always rhombuses?

   A. squares

   B. rectangles

   C. trapezoids

   D. kites

Try This

A **rectangle** is a parallelogram that has all right angles. Which of the following
are rectangles? Write *always, sometimes,* or *never* to complete each sentence.
Explain your answers.

6. Squares are _____ rectangles. Explain. _____

   _____

7. Rhombuses are _____ rectangles. Explain. _____

   _____

8. Trapezoids are _____ rectangles. Explain.

   _____

9. A kite is _____ a parallelogram. Explain. _____

   _____

LESSON 1·5   **What Is a Polygon?**

These are polygons.

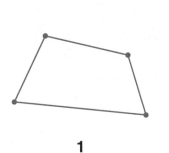      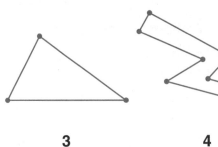   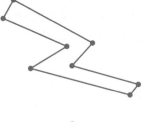

    1   2   3   4

These are NOT polygons.

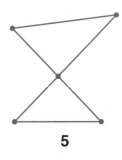   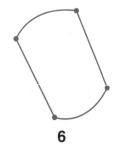   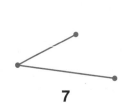      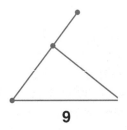

    5   6   7   8   9

1. If you had to explain what a polygon is, what would you say? (*Think:* What do Polygons 1–4 have in common? How are Shapes 5–9 different from Polygons 1–4?)

_____

_____

_____

_____

_____

2. Choose one of the shapes from above. Explain why the shape is not a polygon.

_____

_____

_____

_____

12

Math Boxes

1. Subtract mentally.

    a. $10 - 7 =$ _____

    b. $11 - 5 =$ _____

    c. $14 - 6 =$ _____

    d. $150 - 80 =$ _____

    e. $130 - 70 =$ _____

    f. $160 - 90 =$ _____

2. Draw $\angle MRT$.

    T

    M •

    •
    R

What is another name for $\angle MRT$?

_____

3. Draw and label line segment *AB*.

What is another name for $\overline{AB}$?

_____

4. Name as many rays as you can in the figure below.

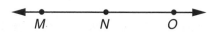

    M   *N*   *O*

Write their names.

_____

_____

_____

5. Which polygons have 2 pairs of parallel sides? Circle the best answer.

    A. square and trapezoid

    B. rectangle and rhombus

    C. triangle and parallelogram

    D. pentagon and square

6. Put these numbers in order from least to greatest.

    10,005   51,000

    5,100   10,500

    _____

    _____

    _____

    _____

LESSON 1·6   # An Inscribed Square

SRB
116

Follow the directions below to make a square that you will tape on the next page.

Step 1 Use your compass to draw a circle on a sheet of colored paper. The circle should be small enough to fit on the next page. Cut out the circle.

Step 2 With your pencil, make a dot in the center of the circle, where the hole is, on both the front and the back.

Step 3 Fold the circle in half. Make sure that the edges match and that the fold line passes through the center. Be sure to make sharp creases.

Step 4 Fold the folded circle in half again so that the edges match.

Step 5 Unfold your circle. The folds should pass through the center of the circle and form 4 right angles.

Step 6 Using a straightedge, connect the endpoints of the folds at the edge of the circle to make a square. Cut out the square.

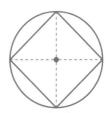

LESSON 1·6 An Inscribed Square *continued*

Now use your compass to find out whether the 4 sides of your square are about the same length.

Place the anchor on one endpoint of a side and the pencil point on the other endpoint of the side. Then, *without changing the compass opening,* try to place the anchor and pencil point on the endpoints of each of the other sides.

If the sides of your square are about the same length, tape the square in the space below. If not, follow the directions on page 14 again. Tape your best square in the space below.

LESSON 1·6   Math Boxes

1. Subtract mentally.

  a. 11 − 2 = _____

  b. 110 − 20 = _____

  c. _____ = 12 − 4

  d. 120 − 40 = _____

  e. _____ = 120 − 90

  f. 160 − 80 = _____

2. Which of the shape(s) below are NOT polygons? _____

A   B   C

3. Draw a quadrangle with only 1 right angle. Draw in the right angle symbol.

How do you know it is a right angle?

_____

93 99

4. Circle the convex polygon(s).

5. Draw and label ray *HA*. Draw point *T* on it.

What is another name for $\overrightarrow{HA}$? _____

91

6. In the numeral 42,318, the 2 stands for 2,000.

  a. The 1 stands for _____.

  b. The 8 stands for _____.

  c. The 4 stands for _____.

  d. The 3 stands for _____.

4

LESSON 1·7   Circle Constructions

Do each of the following 3 constructions on a separate sheet of paper. Try and try again until you are satisfied with your work. Then cut out your 3 best constructions and tape them in your journal.

1. Use your compass to draw a picture of a circular dartboard. It is not necessary to include the details of the board. Tape your best work in the space below. The circles in the dartboard and in your picture are called **concentric circles**.

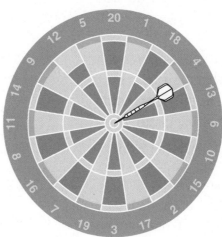

LESSON 1·7 Circle Constructions *continued*

2. **a.** Make a dot near the center of your paper. Use your compass to draw a circle with that dot as its center.

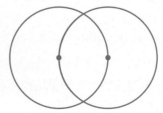

b. *Without changing the opening of your compass,* draw a **congruent** circle that **intersects** the center of the first circle. Mark the center of the second circle.

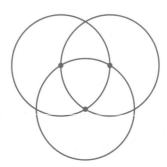

c. *Without changing the opening of your compass,* draw a third congruent circle that intersects the center of each of the first 2 circles.

Try and try again until you are satisfied with your work. Then cut out your circle design and tape it in the space below.

LESSON 1·7 **Circle Constructions** *continued*

Try This

3. Draw this design with your compass. Work on separate sheets of paper until you are satisfied with your work. Color your best design. Then cut it out and tape it in the space below.

   Hint: Start by making the 3-circle design on page 18. Then add more circles to it.

Math Boxes

---

1. Subtract mentally.

  a. 9 − 7 = _____

  b. 10 − 6 = _____

  c. _____ = 16 − 8

  d. 170 − 70 = _____

  e. _____ = 130 − 70

  f. 150 − 90 = _____

2. Draw ∠*TIF*. What is the vertex of ∠*TIF*?

Point _____

F•

          •*T*

I•

---

3. Draw and label line segment *GP*.

What is another name for $\overline{GP}$?

_____

4. Name as many rays as you can in the figure below.

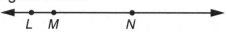

    L *M*   *N*

Write their names.

_____

_____

---

5. Draw a quadrangle with 1 pair of parallel sides.

What kind of quadrangle is this?

_____

6. Put these numbers in order from least to greatest.

  32,000   3,200

  23,000   2,300

  _____

  _____

  _____

  _____

Copying a Line Segment

Steps 1–4 below show you how to copy a line segment.

Step 1 You are given line segment *AB* to copy.

A   B

Step 2 Draw a line segment that is longer than line segment *AB*. Label one of its endpoints *C*.

C

Step 3 Open your compass so that the anchor is on one endpoint of line segment *AB* and the pencil point is on the other endpoint.

A   B

Step 4 *Without changing the compass opening,* place the anchor on point *C* on your second line segment. Make a mark that crosses this line segment. Label the point where the mark crosses the line segment with the letter *D*.

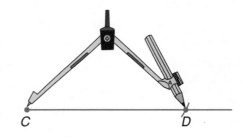

C   D

Line segment *CD* should be about the same length as line segment *AB*. Line segments *CD* and *AB* are **congruent.**

Use a compass and straightedge to copy the line segments shown below. For each problem, begin by drawing a line segment that is longer than the one given.

1.
E   F

2.
M   N

Hexagons in Our World

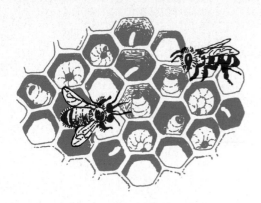

Hexagons are seen in the natural world and in things that people make and use. For example, bees make honeycombs with hexagonal shapes, and snowflakes suggest the shape of a hexagon.

Soccer balls are made up of regular hexagons and regular pentagons.

For many centuries, wonderful tile designs have been created all over the world, especially in Islamic art. As these pictures show, tile designs can be developed in many ways from a pattern that uses hexagons.

Many quilt and fabric designs come from dividing regular hexagons into triangles or rhombuses. You may have made designs like these with pattern blocks. Coloring a design often makes the design more interesting.

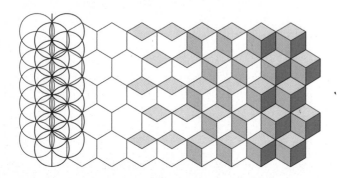

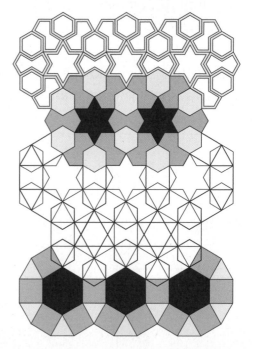

LESSON 1·8 Constructing an Inscribed, Regular Hexagon

Follow each step below. Draw on a separate sheet of paper. Repeat these steps several times. Cut out your best work, and tape it onto the bottom of this page.

Step 1 Draw a circle. (Keep the same compass opening for Steps 2 and 3.) Draw a dot on the circle. Place the anchor of your compass on the dot and make a mark on the circle.

Step 2 Place the anchor of your compass on the mark you just made and make another mark on the circle.

Step 3 Do this four more times to divide the circle into 6 equal parts. The 6th mark should be on the dot you started with or very close to it.

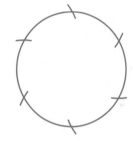

Step 4 With your straightedge, connect the 6 marks on the circle to form a regular hexagon. Use your compass to check that the sides of the hexagon are all about the same length.

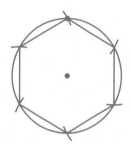

The hexagon is **inscribed** in the circle because each vertex of the hexagon is on the circle.

LESSON 1·8   **More Constructions**

Construct a regular hexagon on a separate sheet of paper. Then divide the hexagon into 6 equilateral triangles. Use your compass to check that the sides of the 6 equilateral triangles are all about the same length.

Try this several times until you are satisfied with your work. Then cut out your best work and tape it in the space below.

LESSON 1·8 Definition Match

Match each description of a geometric figure in Column I with its name
in Column II. Some of the items in Column II do not have a match.

SRB
94–100

I	II

a. a polygon with 4 right angles and
4 sides of the same length

b. a polygon with 4 sides, none of
which are the same length

c. a quadrilateral with exactly 1 pair
of opposite sides that is parallel

d. lines that never intersect

e. a parallelogram with all sides the
same length, but not a rectangle

f. a polygon with 8 sides

g. a polygon with 5 sides

h. an angle that measures 90°

i. a triangle with all sides the same
length

_____ octagon

_____ rhombus

_____ right angle

_____ trapezoid

_____ hexagon

_____ square

_____ equilateral triangle

_____ perpendicular lines

_____ parallel lines

_____ pentagon

_____ isosceles triangle

_____ quadrangle

LESSON 1·8 **Math Boxes**

---

1. Subtract mentally.

   a. 14 − 9 = _____

   b. 140 − 90 = _____

   c. _____ = 18 − 9

   d. 180 − 90 = _____

   e. _____ = 110 − 70

   f. _____ = 150 − 60

---

2. Which of the shape(s) below are polygons? _____

    A   B   C

SRB 96

---

3. Draw a quadrangle that has 2 pairs of parallel sides and no right angles.

What kind of quadrangle is this?

_____

SRB 99 100

---

4. Circle the concave (nonconvex) polygon(s).

SRB 97

---

5. Draw and label ray *CA*.
Draw point *R* on it.

What is another name for ray *CA*?

_____

SRB 91

---

6. In the numeral 30,516, what does the 3 stand for? Circle the best answer.

   A. 3,000

   B. 30

   C. 30,000

   D. 300,000

SRB 4

Math Boxes

1. Add.

    a.   64
         + 32

    b.   48
         + 96

SRB 10 11

2. Subtract.

    a.   78
         − 42

    b.   81
         − 36

SRB 12–15

3. Put these numbers in order from least to greatest.

    46,000   40,600

    4,600   4,006

    _____

    _____

    _____

    _____

SRB 4

4. In the numeral 78,965,

    a. the 8 stands for _____.

    b. the 6 stands for _____.

    c. the 7 stands for _____.

    d. the 9 stands for _____.

SRB 4

5. Use the following list of numbers to answer the questions:

    12, 3, 15, 6, 12, 14, 6, 5, 9, 12

    a. Which number is the least? _____

    b. Which number is the greatest? _____

    c. What is the difference between the least and greatest numbers? _____

    d. Which number appears most often? _____

SRB 73

LESSON 2·1 A Visit to Washington, D.C.

Refer to pages 267–269 in your *Student Reference Book.*

1. About how many people tour the White House every year?
   Check the best answer.

   —— between 100 thousand and 1 million ____ between 1 million and 10 million

   —— between 10 million and 100 million —— between 100 million and 1 billion

2. About how many people ride the Washington Metrorail on an average weekday?
   Check the best answer.

   ____ between 100 thousand and 1 million —— between 1 million and 10 million

   ____ between 10 million and 100 million —— between 100 million and 1 billion

3. The Library of Congress adds about _____ items each day. About how many

   days does it take to add 50,000 items to the Library of Congress? _____

4. Write the year that each event happened. Then draw a dot for each event
   on the timeline below. Label the dot with the correct letter and date.

   A The year the Metrorail opened ____*1976*____

   B The year of the flight of the *Flyer* _____

   C The year the Washington Monument was completed _____

   D The year the Lincoln Memorial was dedicated _____

   E The year of the first nonstop flight across the Atlantic _____

   F The year the Jefferson Memorial was dedicated _____

   G The year of the first landing on the moon _____

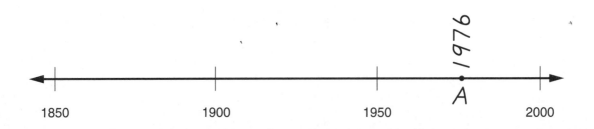

LESSON 2·1   **Math Boxes**

---

1. Add mentally.

a. $3 + 5 =$ _____

b. $30 + 50 =$ _____

c. $300 + 500 =$ _____

d.   _____ $= 9 + 7$

e.   _____ $= 90 + 70$

f.   _____ $= 900 + 700$

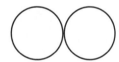
SRB
10 11

---

2. What is the value of the digit 5
in **5**60? ____500____

What is the value of the digit 7
in the numbers below?

a.   474 _____

b.   70,158 _____

c.   187,943 _____

d.   2,731,008 _____

SRB
4

---

3. Circle the pair of concentric circles.

---

4. I am a 2-dimensional figure.
I have two pairs of parallel sides.
All my angles have the same measure.
All my sides are the same length.

What am I? _____

Use your Geometry Template to draw me.

SRB
100

---

5. A giant tortoise can live for about 150
years. An elephant can live for about 78
years. About how much longer can a giant
tortoise live than an elephant? Fill in the
circle next to the best answer.

Ⓐ 88 years

Ⓑ 238 years

Ⓒ 72 years

Ⓓ 228 years

SRB
178

---

6. Multiply mentally.

a. $2 \times 1 =$ _____

b.   _____ $= 5 \times 0$

c.   _____ $= 5 \times 2$

d. $5 \times 4 =$ _____

e. $3 \times 10 =$ _____

SRB
16

LESSON
2·2

Name-Collection Boxes

Write five names in each box below. Use as many different kinds of numbers (such as counting numbers, fractions, decimals, negative numbers) and different operations (+, −, ×, ÷) as you can. Draw a star next to the name you find most interesting.

SRB
149

1.

16
8 × 2
970 − 954
10 + (60 ÷ 10)

2.
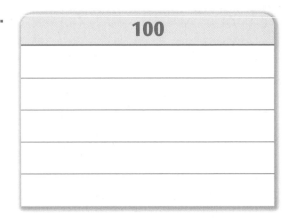

24

3.

50

4.

100

Make up your own name-collection boxes. Use different kinds of numbers and operations.

5.

6.

LESSON 2·2 **Math Boxes**

---

1. a. Write the largest number you can make with the digits 5, 2, 3, 0, 6, 0. Use each digit only once.

_____

b. Use the same digits and write the smallest number you can make. Do not start with 0.

_____

SRB
4

2. Add mentally or with a paper-and-pencil algorithm.

a.   37
      + 142

b.   468
      + 394

SRB
10 11

---

3. Draw a convex polygon.

SRB
97

4. Measure these line segments to the nearest centimeter.

a. ▬▬▬▬▬▬▬▬▬▬▬▬

About _____ centimeters

b. ▬▬▬▬▬▬▬

About _____ centimeters

SRB
128

---

5. Complete.

a. 4 ft = _____ in.

b. 4 ft = _____ yd _____ in.

c. 2 yd = _____ ft

d. 72 in. = _____ yd _____ ft

e. 6,756 in. = _____ ft

SRB
129

6. Divide mentally.

a. $9 \div 9 =$ _____

b. _____ $= 12 \div 2$

c. _____ $= 20 \div 5$

d. $40 \div 10 =$ _____

e. $14 \div 7 =$ _____

SRB
20

LESSON 2·3   **Place-Value Chart**

Number	Hundred Millions	Ten Millions	Millions	Hundred Thousands	Ten Thousands	Thousands	Hundreds	Tens	Ones
	100M	10M	M	100K	10K	K	H	T	O

LESSON 2·3   Taking Apart, Putting Together

Complete.

1. In 574,

   5 is worth _____ 500 _____

   7 is worth _____

   4 is worth _____

2. In 9,027,

   9 is worth _____

   0 is worth _____

   2 is worth _____

3. In 280,743,

   8 is worth _____

   2 is worth _____

   4 is worth _____

4. In 56,010,837,

   6 is worth _____

   1 is worth _____

   5 is worth _____

5. In 705,622,463,

   5 is worth _____

   6 is worth _____

   7 is worth _____

6. In 123,456,789,

   4 is worth _____

   3 is worth _____

   2 is worth _____

Add.

7.   900
    70
+   5

8. 30,000
  7,000
    50
+   2

9. 50,000,000
  9,000,000
    60,000
     2,000
      800
+   50

10. 300,000,000
   9,000,000
    200,000
     70,000
       30
+   1

11. Think about why we need zeros when writing numbers. What would happen if you did not write the zero in the number 5,074?

_____

_____

LESSON 2·3   **Polygon Checklist**

Place a check mark next to all of the statements that are true about each figure.
Write an additional true statement for each figure.

1.

_____ 1 pair of parallel sides

_____ at least 1 right angle

_____ quadrangle

_____ polygon

_____ concave

_____ parallelogram

✓ _____

2.

_____ 4 sides of equal length

_____ kite

_____ square

_____ parallelogram

_____ convex

_____ opposite sides parallel

✓ _____

3.

_____ all sides of equal length

_____ all angles of equal measure

_____ one right angle

_____ polygon

_____ equilateral triangle

_____ 1 pair of parallel sides

✓ _____

4.

_____ regular polygon

_____ all sides of equal length

_____ all angles of equal measure

_____ pentagon

_____ octagon

_____ all angles smaller than right angles

✓ _____

Math Boxes

---

1. Add mentally.

  a. 4 + 5 = _____

  b. 40 + 50 = _____

  c. 400 + 500 = _____

  d. _____ = 5 + 8

  e. _____ = 50 + 80

  f. _____ = 500 + 800

2. What is the value of the digit 8 in the numbers below?

  a. 5**8**4 _____

  b. 3**8**,067 _____

  c. 49,**8**41 _____

  d. **8**20,731 _____

  e. **8**,391,467 _____

---

3. Use your compass to draw a pair of concentric circles.

4. I am a 2-dimensional figure.
I have two pairs of parallel sides.
None of my angles is a right angle.
All of my sides are the same length.

What am I? _____

Use your Geometry Template to draw me.

---

5. A sailfish can swim at a speed of 110 kilometers per hour. A tiger shark can swim at a speed of 53 kilometers per hour. How much faster can a sailfish swim than a tiger shark?

_____ kilometers per hour

6. Multiply mentally.

  a. 8 × 1 = _____

  b. _____ = 9 × 0

  c. _____ = 5 × 6

  d. 5 × 5 = _____

  e. 7 × 10 = _____

LESSON 2·4 Calculator "Change" Problems

1. Follow your teacher's directions to complete the "change" problems below.
   Use your calculator.

	Start with	Place of Digit	Change to	Operation	New Number
a.	570	Tens			
b.	409	Hundreds			
c.	54,463	Thousands			
d.	760,837	Tens			
e.	52,036,458	Ones			
f.		Ten Thousands			
g.		Millions			

2. Complete these calculator "change" problems on your own.

	Start with	Place of Digit	Change to	Operation	New Number
a.	893	Tens	3	−	
b.	5,489	Hundreds	7	+	
c.	94,732	Thousands	6	+	
d.	218,149	Ten Thousands	0	−	
e.	65,307,000	Millions	9	+	
f.	873,562,003	Ten Millions	1		
g.	103,070,651	Hundred Millions	8		

LESSON 2·4

Math Boxes

---

1. What is the largest number you can make with the digits 3, 0, 3, 8, and 0? Fill in the circle next to the best answer.

(A) 83,003

(B) 83,030

(C) 83,300

(D) 80,033

SRB
4

2. Add mentally or with a paper-and-pencil algorithm.

a.   145
   + 34

b.   297
   +136

SRB
10 11

---

3. Draw a concave pentagon.

SRB
97

4. Measure these line segments to the nearest centimeter.

a. _____

About _____ centimeters

b. _____

About _____ centimeters

SRB
128

---

5. Complete.

a. 14 in. = _____ ft _____ in.

b. _____ in. = 2 ft

c. _____ ft = 7 yd

d. 1 yd 1 ft = _____ in.

e. 413 ft = _____ yd _____ ft

SRB
129

6. Divide mentally.

a. 16 ÷ 2 = _____

b. 20 ÷ 10 = _____

c. _____ = 40 ÷ 5

d. 60 ÷ 10 = _____

e. _____ = 45 ÷ 5

SRB
20

**LESSON
2·5**   **Counting Raisins**
                                                       71–75

1. Use your $\frac{1}{2}$-ounce box of raisins. Complete each step when the teacher tells you. Stop after you complete each step.

2. Make a tally chart of the class data.

Number of Raisins	Number of Boxes

   a. Don't open your box yet. **Guess** about how many raisins are in the box.

      About _____ raisins

   b. Open the box. Count the number of raisins in the top layer. Then **estimate** the total number of raisins in the box.

      About _____ raisins

   c. Now **count** the raisins in the box.

      How many? _____ raisins

3. Find the following **landmarks** for the class data.

   a. What is the **maximum,** or largest, number of raisins found? _____

   b. What is the **minimum,** or smallest, number of raisins found? _____

   c. What is the **range?** (Subtract the minimum from the maximum.) _____

   d. What is the **mode,** or most frequent number of raisins found? _____

Try This

4. What is the **median** number of raisins found? _____

5. What is the **mean** number of raisins found? _____

Math Boxes

---

1. A number has

    6 in the hundreds place,
    1 in the millions place,
    2 in the tens place,
    8 in the hundred-thousands place,
    5 in the ones place,
    3 in the thousands place, and
    4 in the ten-thousands place.

Write the number.

___ , ___ ___ ___ , ___ ___ ___

SRB
4

---

2. Write five names for 34.

34

SRB
149

---

3. Write >, <, or = to make each number sentence true.

  a. 14 _____ 26

  b. 3,003 _____ 3,300

  c. 12 + 12 _____ 24

  d. 200 − 50 _____ 100

  e. 30 + 30 _____ 50 + 10

SRB
148 149

---

4. Name the two pairs of parallel sides in parallelogram *HIJK*.

_____ and _____

_____ and _____

SRB
94

---

5. Measure these line segments to the nearest $\frac{1}{2}$ centimeter.

  a. _____

    About _____ centimeters

  b. _____

    About _____ centimeters

SRB
128

---

6. Multiply mentally.

  a. 5 × 7 = _____

  b. 3 × _____ = 18

  c. _____ × 7 = 56

  d. 9 × _____ = 45

  e. 8 × 4 = _____

SRB
16

LESSON 2·6   **Family Size**

Follow your teacher's directions and complete each step.

1. How many people are in your family? _____ people
    Write the number on a stick-on note.

2. Make a line plot of the family-size data for the class.
    Use **X**s in place of stick-on notes.

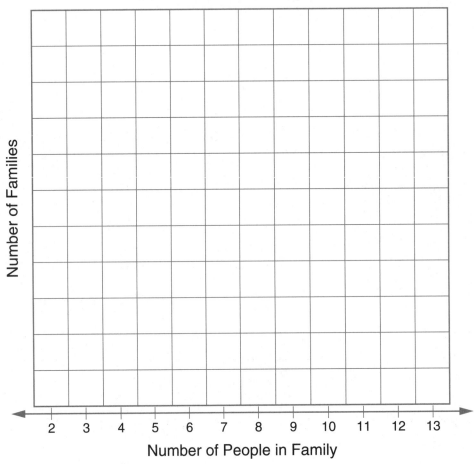

Class Data on Family Size

Number of Families

2   3   4   5   6   7   8   9   10   11   12   13

Number of People in Family

3. Find the following landmarks for the class data:

    a. What is the **maximum** (largest) number of people in a family? _____ people

    b. What is the **minimum** (smallest) number of people in a family? _____ people

    c. What is the **range?** (Subtract the minimum from the maximum.) _____ people

    d. What is the **mode** (most frequent family size)? _____ people

4. What is the **median** family size for the class? _____ people

LESSON 2·6

Math Boxes

---

1. Add mentally.

   a. $2 + 7 =$ _____

   b. $20 + 70 =$ _____

   c. $200 + 700 =$ _____

   d. _____ $= 8 + 4$

   e. _____ $= 80 + 40$

   f. _____ $= 800 + 400$

SRB
10 11

---

2. Find the median of the data set.

       3, 45, 13, 15, 3, 7, 19

Fill in the circle next to the best answer.

  Ⓐ 3

  Ⓑ 13

  Ⓒ 15

  Ⓓ 45

SRB
73

---

3. Subtract mentally or with a paper-and-pencil algorithm.

  a. $\begin{array}{r} 147 \\ -\ 56 \\ \hline \end{array}$ **b.** $\begin{array}{r} 531 \\ -\ 246 \\ \hline \end{array}$

SRB
12–15

---

4. Write 4,007,392 in words.

_____

_____

_____

_____

_____

SRB
4

---

5. A royal python can be 35 feet long. An anaconda can be 28 feet long. What would be their combined length, end-to-end?

_____ feet

---

6. Tell whether each number sentence is true or false.

  a. $14 + 7 = 22$ _____

  b. $36 = 15 + 5$ _____

  c. $45 - 12 = 33$ _____

  d. $27 = 40 - 13$ _____

SRB
148

LESSON 2·7 **Partial-Sums Addition**

Write a number model for a ballpark estimate. Solve Problems 1–3 using the partial-sums method. Solve Problems 4–6 using any method. Compare your answer with your estimate to see if your answer makes sense.

1. $\begin{array}{r} 76 \\ +\ 38 \\ \hline \end{array}$	**2.** $\begin{array}{r} 647 \\ +\ 936 \\ \hline \end{array}$	**3.** $\begin{array}{r} 1,672 \\ +\ 3,221 \\ \hline \end{array}$
Ballpark estimate: _____	Ballpark estimate: _____	Ballpark estimate: _____
4. $\begin{array}{r} 66 \\ +\ 28 \\ \hline \end{array}$	**5.** $\begin{array}{r} 736 \\ +\ 645 \\ \hline \end{array}$	**6.** $\begin{array}{r} 7,854 \\ +\ 4,550 \\ \hline \end{array}$
Ballpark estimate: _____	Ballpark estimate: _____	Ballpark estimate: _____

Try This

7. Name three 4-digit numbers whose sum is 17,491.

_____ + _____ + _____ = 17,491

LESSON 2·7 Column Addition

Write a number model for a ballpark estimate. Solve Problems 1–3 using the column-addition method. Solve Problems 4–6 using any method. Compare your answer with your estimate to see if your answer makes sense.

SRB 11

1. $\begin{array}{r} 94 \\ + 47 \\ \hline \end{array}$ Ballpark estimate: _____	**2.** $\begin{array}{r} 385 \\ + 726 \\ \hline \end{array}$ Ballpark estimate: _____	**3.** $\begin{array}{r} 2{,}538 \\ + 4{,}179 \\ \hline \end{array}$ Ballpark estimate: _____
4. $\begin{array}{r} 49 \\ + 33 \\ \hline \end{array}$ Ballpark estimate: _____	**5.** $\begin{array}{r} 469 \\ + 946 \\ \hline \end{array}$ Ballpark estimate: _____	**6.** $\begin{array}{r} 4{,}614 \\ + 6{,}058 \\ \hline \end{array}$ Ballpark estimate: _____

Try This

7. Name four 4-digit numbers whose sum is 15,706.

_____ + _____ + _____ + _____ = 15,706

LESSON 2·7   **Math Boxes**

---

1. A number has

3 in the millions place,
1 in the ones place,
8 in the thousands place,
9 in the ten-thousands place,
0 in the tens place,
6 in the hundred-thousands place, and
5 in the hundreds place.

Write the number.

___ , ___ ___ ___ , ___ ___ ___

SRB
4

2. Write five names for 100.

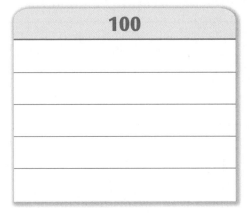

100

SRB
149

---

3. Write >, <, or = to make each number sentence true.

a. 16 + 11 _____ 47

b. 206 _____ 602

c. 150 − 50 _____ 100

d. 62 + 10 + 10 _____ 62 − 10 − 10

e. 423,726 _____ 413,999

SRB
148 149

4. Draw a parallelogram. Label the vertices so that side *AB* is parallel to side *CD*.

SRB
99 100

---

5. Measure these line segments to the nearest $\frac{1}{2}$ centimeter.

a. _____

About _____ cm

b. _____

About _____ cm

SRB
128

6. Multiply mentally.

a. 5 × 8 = _____

b. 2 × _____ = 16

c. 7 × _____ = 21

d. _____ × 9 = 54

e. 8 × 3 = _____

SRB
16

LESSON 2·8   **Math Boxes**

---

1. Add mentally.

a. $2 + 4 =$ _____

b. $20 + 40 =$ _____

c. $200 + 400 =$ _____

d. _____ $= 8 + 6$

e. _____ $= 80 + 60$

f. _____ $= 800 + 600$

SRB
10 11

---

2. Find the following landmarks for this set of numbers: 12, 16, 23, 15, 16, 19, 18.

a. median _____

b. mode _____

c. maximum _____

d. minimum _____

e. range _____

f. mean _____

SRB
73–75

---

3. Subtract mentally or with a paper-and-pencil algorithm.

a.   231
      − 84
      ‾‾‾‾‾

b.   603
      − 466
      ‾‾‾‾‾

SRB
12–15

---

4. Write 8,042,176 in words.

_____

_____

_____

_____

_____

SRB
4

---

5. An ostrich can weigh about 345 pounds. An emu can weigh about 88 pounds. How much would they weigh together?

_____ pounds

---

6. Tell whether each number sentence is true or false.

a. $18 + 9 = 37$ _____

b. $29 = 17 + 12$ _____

c. $42 − 15 = 27$ _____

d. $17 = 40 − 24$ _____

e. $154 − 65 = 99$ _____

SRB
148

LESSON 2·8   Head Sizes

Ms. Woods owns a clothing store. She is trying to decide how many children's hats to stock in each possible size. Should she stock the same number of hats in each size? Or should she stock more hats in some sizes and fewer in others?

Help Ms. Woods decide. Pretend that she has asked each class in your school to collect and organize data about students' head sizes. She plans to combine the data and then use it to figure out how many hats of each size to stock.

As a class, collect and organize data about one another's head sizes.

1. Ask your partner to help you measure the distance around your head.

   ◆ Wrap the tape measure once around your head.

   ◆ See where the tape touches the end tip of the tape measure.

   ◆ Read the mark where the tape touches the end tip.

   ◆ Read this length to the nearest $\frac{1}{2}$ centimeter.

   Record your head size.   About _____ cm

2. What is the median head size for the class?   About _____ cm

3. Find the following landmarks for the head-size data shown in the bar graph on journal page 47.

   Minimum: _____   Maximum: _____   Range: _____

   Mode: _____   Median: _____

4. How would the landmarks above help Ms. Woods, a clothing store owner, decide how many baseball caps of each size to stock?

   _____

   _____

   _____

   _____

LESSON 2·8 **Head Sizes** *continued*

Make a bar graph of the head-size data for the class.

SRB 76

title

label

label

LESSON 2·8 **Head Sizes** *continued*

1. Count the Xs in the line plot you created on journal page 47B to answer the questions below.

   a. What is the largest head size? _____

   b. What is the smallest head size? _____

   c. What is the difference between the largest and the smallest head size? Write a number model to show how you found your answer.

   _____

   d. What is the mode of the head-size data? _____

   e. What is the difference between your head size and the mode? Write a number model to show how you found your answer.

   _____

2. Did anything about the data surprise you? Did you expect the head sizes to be close together or spread out?

   _____

   _____

   _____

3. The average head size for an infant at birth is 35 cm. Use a number model to show how your head size compares to that of an infant.

   _____

4. Imagine that you measured the head sizes of a room full of newborn babies and created a line plot. What would you expect the graph to look like? Explain your answer.

   _____

   _____

   _____

LESSON 2·8 Head Sizes *continued*

Make a line plot of the head-size data for the class.

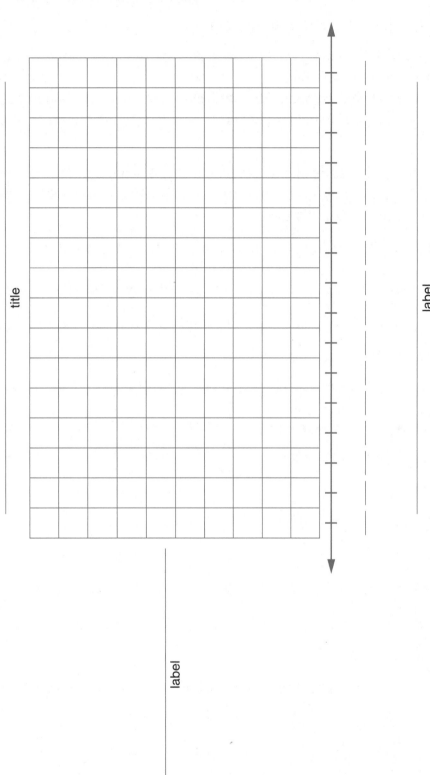

title

label

label

LESSON 2·8   How to Construct a Kite

Follow the steps to construct a kite in the space below.

Step 1   Draw points *A* and *B*. (*See example below.*)

Step 2   Set your compass opening so that it is a little more than half the distance between *A* and *B*. Place the point of the compass on *A* and draw an arc. *Without changing the compass opening,* place the point of the compass on *B* and draw a second arc that intersects the first arc. Label the point where the two arcs meet *C*.

Step 3   Change your compass opening. Set it so that it is almost the full distance between *A* and *B* (about $\frac{4}{5}$ of the width). Repeat Step 2 as shown in the picture, and label the new point of intersection *D*.

Step 4   With your straightedge, connect the 4 points to form a quadrangle.

Example:

Draw your kite here.

D
×

A •   • B

×
C

Steps 1–3

D
×

A •   • B

×
C

Step 4

LESSON 2·9 Trade-First Subtraction

Write a number model for a ballpark estimate. Solve Problems 1–3 using the trade-first method. Solve Problems 4–6 using any method. Use your ballpark estimates to see if your answers make sense.

1.

```
    58
 −  39
```


Ballpark estimate:

_____

2.

```
   600
 − 379
```


Ballpark estimate:

_____

3.

```
   2,936
 − 1,657
```


Ballpark estimate:

_____

4.

```
    73
 −  37
```


Ballpark estimate:

_____

5.

```
   900
 − 461
```


Ballpark estimate:

_____

6.

```
   2,468
 − 1,789
```


Ballpark estimate:

_____

Try This

7. Name two 3-digit numbers whose difference is 357.

_____ − _____ = 357

LESSON 2·9   Partial-Differences Subtraction

Write a number model for a ballpark estimate. Solve Problems 1–3 using the partial-differences method. Solve Problems 4–6 using any method. Use your ballpark estimates to see if your answers make sense.

1.
$$\begin{array}{r} 86 \\ -\ 37 \\ \hline \end{array}$$

Ballpark estimate:

_____

2.
$$\begin{array}{r} 900 \\ -\ 485 \\ \hline \end{array}$$

Ballpark estimate:

_____

3.
$$\begin{array}{r} 7,584 \\ -\ 2,806 \\ \hline \end{array}$$

Ballpark estimate:

_____

4.
$$\begin{array}{r} 61 \\ -\ 26 \\ \hline \end{array}$$

Ballpark estimate:

_____

5.
$$\begin{array}{r} 400 \\ -\ 271 \\ \hline \end{array}$$

Ballpark estimate:

_____

6.
$$\begin{array}{r} 3,681 \\ -\ 1,803 \\ \hline \end{array}$$

Ballpark estimate:

_____

Try This

7. Name two 4-digit numbers whose difference is 4,203.

_____ – _____ = 4,203

LESSON 2·9   **Math Boxes**

---

1. A number has

6 in the tens place,
9 in the millions place,
4 in the thousands place,
3 in the ten-thousands place,
2 in the hundred-thousands place,
0 in the ones place, and
8 in the hundreds place.

Write the number.

___, ___ ___ ___, ___ ___ ___
SRB 4

---

2. Write five names for 1,000.

1,000

SRB 149

---

3. Which number sentence is true? Fill in the circle next to the best answer.

Ⓐ 34 − 4 = 31

Ⓑ 812 < 218

Ⓒ 423 + 20 > 443

Ⓓ 123 + 5 + 5 = 113 + 20

SRB 148 149

---

4. Draw a polygon that has no parallel sides.

SRB 94

---

5. Measure these line segments to the nearest $\frac{1}{2}$ centimeter.

a. _____

About _____ cm

b. _____

About _____ cm

SRB 128

---

6. Multiply mentally.

a. _____ = 5 × 4

b. 5 × 6 = _____

c. 9 × 3 = _____

d. 4 × _____ = 24

e. 6 × _____ = 48

SRB 16

LESSON 2·10   **Math Boxes**

1. Divide mentally.

   a. 35 ÷ 5 = _____

   b. 56 ÷ _____ = 8

   c. 32 ÷ _____ = 4

   d. 24 ÷ _____ = 6

   e. 72 ÷ 8 = _____

   f. 40 ÷ 5 = _____

20

2. Multiply mentally.

   a. 4 × 7 = _____

   b. 3 × _____ = 15

   c. 7 × _____ = 42

   d. 9 × _____ = 36

   e. 6 × 0 = _____

   f. 1 × 9 = _____

16

3. Complete the square facts.

   a. 64 ÷ 8 = _____

   b. 49 ÷ 7 = _____

   c. _____ = 4 × 4

   d. _____ = 3 × 3

   e. 25 ÷ 5 = _____

158

4. Tell whether each number sentence is true or false.

   a. 46 + 12 = 53 _____

   b. 36 = 22 + 14 _____

   c. 13 = 84 − 71 _____

   d. 52 − 20 = 34 _____

148

5. A grizzly bear can weigh 786 pounds. An American black bear can weigh 227 pounds. What is their combined weight?

_____ pounds

6. On average, India produces about 855 movies per year. The United States produces about 762 movies. On average, how many fewer movies per year does the United States produce than India?

_____

52

Date _____   Time _____

"What's My Rule?"

Complete the "What's My Rule?" tables and state the rules.

1.

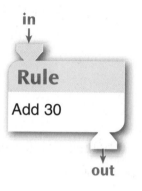

in
Rule
Add 30
out

in	out
30	
80	
20	
150	
290	

2.

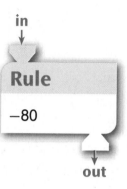

in
Rule
−80
out

in	out
	50
	210
	20
	270
	340

3.

in
Rule
out

in	out
49	72
151	
272	295
	611
	503

4. Rule: There are 12 inches in 1 foot.

in	out
3	36
	60
10	
	264
	720

Try This

5. Rule: _____

in	out
17	−8
12	
27	2
−5	
	0

6. Create your own.

Rule: _____

in	out

A Polygon Alphabet

Try reading this message:

ALL OF THESE LETTERS ARE POLYGONS.

1. Use a straightedge to design a polygon letter for each of the letters shown below. You'll have to simplify, because a polygon can't have any curves, and it can't have any "holes."

   For example, if you look at the letter "P," you see that there is no opening in the upper part. Making it look like this, ▷, would make it easier to read, but it would not be a polygon.

B	C	D
F	M	X

2. Which of the letters you drew are nonconvex (concave) polygons? _____
   How do you know?

   _____

3. Do any of the letters you drew have special names as polygons? Explain.

   _____

   _____

Try This

4. On a separate sheet of paper, design polygon letters for the rest of the uppercase (capital) letters in the alphabet, the 26 lowercase (small) letters, or the 10 digits (0–9).

LESSON 3·1 Math Boxes

1. Write >, <, or = to make each number sentence true.

a. 1 million _____ 100,000

b. 73,099 _____ 71,999

c. 304,608 _____ 304,809

d. 5,682 _____ 7 hundred

e. 5,000,236 _____ 5,000,099

SRB
6 149

2. Number of spelling words correct for 10 students on the spelling test:

25, 19, 16, 25, 18, 19, 25, 24, 25, 23

a. What is the range for this set of numbers? _____

b. What is the median? _____

SRB
73

3. Make a ballpark estimate. Write a number model to show your strategy.

a. 3,389 + 2,712

_____ + _____ = _____

b. 3,452 − 1,147

_____ − _____ = _____

SRB
181

4. Complete.

a. 21 ft = _____ yd

b. 4 ft = _____ in.

c. 16 ft = _____ yd _____ ft

d. 2 yd 2 ft = _____ in.

e. _____ ft _____ in. = 568 in.

SRB
129

5. Complete.

a. 7, 15, 23, ____, ____, ____

Rule: _____

b. 49, 42, ____, 28, ____, ____

Rule: _____

c. ____, ____, 53, 59, ____, 71

Rule: _____

SRB
160 161

6. Solve mentally or with a paper-and-pencil algorithm.

a. $3.56
   + $2.49
   _____

b. $6.25
   − $5.01
   _____

SRB
34–37

LESSON
3·2

Factor Pairs of Prime Numbers

1. In the table below, list all the factor pairs of each number.

Number	Factor Pairs
2	*1 and 2*
3	
4	
5	
6	*1 and 6 2 and 3*
7	
8	
9	
10	
11	
12	

2. Name a number in the table above that is not a prime number. Explain how you know it is not prime.

_____

_____

3. Name at least three prime numbers that are not in the table above.

_____

4. Choose one of your answers from Problem 3. Explain how you know it is a prime number.

_____

_____

LESSON 3·2 Math Boxes

1. The numbers 28, 35, and 42 are all multiples of ___. Circle the best answer.

A   7

B   4

C   6

D   2

SRB 9

2. Complete the "What's My Rule?" table and state the rule.

Rule: _____

in	out
236	331
682	777
	486
938	
647	

SRB 162–166

3. Earth is covered by a rocky outer layer called the *crust,* which is made up of many elements.

a. Is there more aluminum or silicon in Earth's crust?

_____

b. What percentage of Earth's crust is aluminum?

_____

c. Which element makes up most of Earth's crust?

_____

**Elements Found in Earth's Crust
(percent by weight)**

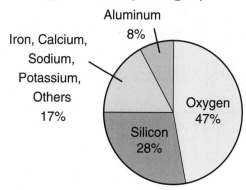

Aluminum 8%

Iron, Calcium, Sodium, Potassium, Others 17%

Oxygen 47%

Silicon 28%

4. Name as many line segments as you can in the figure below.

A   B   C   D

_____

SRB 90

5. Put these numbers in order from smallest to largest.

0.6   0.06   0.43   0.9

_____

_____

_____

_____

SRB 32 33

LESSON
3·3

Patterns in Multiplication Facts

SRB
16

Math Message

Look at the Multiplication/Division Facts Table on the inside front cover of your journal.

1. Find a pattern in the 9s multiplication facts. Describe the pattern.

_____

_____

_____

_____

2. Find a pattern in the 5s multiplication facts. Describe the pattern.

_____

_____

_____

_____

3. What other patterns can you find in the multiplication facts? Write about some of them.

_____

_____

_____

_____

Math Boxes

---

1. Write >, <, or = to make each number sentence true.

a. 45,699 _____ 45,609

b. 67,749 _____ 66,749

c. 208,775 _____ 200 million

d. 1,000,000 _____ 858,192

e. 2 million _____ 20,000,000

 SRB 6 149

---

2. Number of days it took 10 students to complete their science projects:

6, 4, 10, 11, 8, 6, 14, 9, 3, 12

a. What is the range for this set of numbers?

_____

b. What is the median?

_____

 SRB 73

---

3. Make a ballpark estimate. Write a number model to show your strategy.

a. 1,459 + 291

_____ + _____ = _____

b. 1,381 − 646

_____ − _____ = _____

SRB 181

---

4. Complete.

a. 3 yd = _____ ft

b. 4 ft = _____ in.

c. 54 in. = _____ ft _____ in.

d. $\frac{1}{2}$ yd = _____ ft _____ in.

e. $17\frac{1}{2}$ yd = _____ in.

 SRB 129

---

5. Complete.

a. 20, 35, 50, _____, _____, _____

Rule: _____

b. _____, 68, _____, 94, _____, 120

Rule: _____

c. 58, _____, _____, _____, _____, −2

Rule: _____

 SRB 160 161

---

6. Solve mentally or with a paper-and-pencil algorithm.

a. $10.97
   + $15.60
   _____

b. $4.56
   − $2.07
   _____

 SRB 34–37

LESSON 3·4 Math Boxes

1. Complete.

a. Name four multiples of 5.

_____, _____, _____, _____

b. Name four multiples of 9.

_____, _____, _____, _____

SRB
9

2. Complete the "What's My Rule?" table and state the rule.

Rule: _____

in	out
502	346
1,238	1,082
	927
871	
1,600	

SRB
162–166

3. Mr. Rosario's fourth graders collected data about their favorite fruits.

Favorite Fruits

a. Which kind of fruit do most students prefer? _____

b. Which kind is preferred least? _____

c. Do more people prefer apples or peaches? _____

d. About $\frac{1}{4}$ of the students prefer _____ .

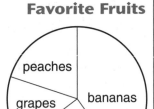

4. Name as many line segments as you can in the figure below.

E   F   G   H

_____

SRB
90

5. Put these numbers in order from smallest to largest.

0.7   0.007   0.5   0.63

_____

_____

_____

_____

SRB
32 33

LESSON 3·5
Multiplication and Division

Equivalents			
$3 * 4$	$12 / 3$	$12 \div 3$	$3 < 5$ (< means "is less than")
3×4	$\dfrac{12}{3}$	$3\overline{)12}$	$5 > 3$ (> means "is greater than")

1. Choose 3 Fact Triangles. Write the fact family for each.

_____ * _____ = _____ | _____ × _____ = _____ | _____ * _____ = _____

_____ * _____ = _____ | _____ × _____ = _____ | _____ * _____ = _____

_____ / _____ = _____ | _____ / _____ = _____ | _____ ÷ _____ = _____

_____ / _____ = _____ | _____ / _____ = _____ | _____ ÷ _____ = _____

2. Solve each division fact.

a. $27 / 3 = $ _____   Think: How many 3s in 27?

b. _____ $= 45 / 5$ Think: 45 is 5 times as many as what number?

c. $36 \div 6 = $ _____   Think: 6 times what number equals 36?

d. $24 / 8 = $ _____   Think: 24 is 8 times as many as what number?

Try This

3. A cashier has 5 rolls of quarters and 6 rolls of dimes in the cash register.
Each roll of quarters is worth $10, and each roll of dimes is worth $5.

a. How much are the rolls of quarters and dimes worth in all? $_____

b. How many quarters are in 1 roll?   _____ quarters

c. How many quarters are in the 5 rolls?   _____ quarters

d. How many dimes are in 1 roll?   _____ dimes

e. How many dimes are in the 6 rolls?   _____ dimes

f. There is also $7.50 worth of half-dollars in
the cash register. How many half-dollars is that?   _____ half-dollars

LESSON 3·5 Math Boxes

1. Write >, <, or = to make each number sentence true.

a. 5,389 _____ 3,389

b. 70,642 _____ 70,699

c. 6 million _____ 6,000,000

d. 8,000,032 _____ 8 million, 32 thousand

e. 400 + 30 + 5 _____ 4,000 + 30 + 5

6 149

2. The number of glasses of milk drunk by 10 students in a week:

16, 13, 15, 20, 8, 10, 15, 12, 10, 18

What is the range? Circle the best answer.

A 8

B 20

C 12

D 14

73

3. Make a ballpark estimate. Write a number model to show your strategy.

a. 13,685 − 8,379

_____ − _____ = _____

b. 7,602 − 3,213

_____ − _____ = _____

181

4. Complete.

a. 31 in. = _____ ft _____ in.

b. 17 ft = _____ yd _____ ft

c. _____ ft = 14 yd

d. _____ in. = 2 yd 1 ft

e. $2\frac{1}{4}$ miles = _____ ft

129

5. Complete.

a. 7, 11, 15, _____, _____, _____

Rule: _____

b. _____, _____, _____, 22, 25, 28

Rule: _____

c. _____, 14, _____, 28, _____, 42

Rule: _____

160 161

6. Solve mentally or with a paper-and-pencil algorithm.

a. $2.27
   + $4.96
  ‾‾‾‾‾‾‾

b. $5.00
   − $3.64
  ‾‾‾‾‾‾‾

34–37

LESSON 3·6 **Math Boxes**

1. Complete the name-collection box.

125

SRB
149

2. A number has

2 in the tens place,
8 in the hundred-thousands place,
5 in the millions place,
7 in the hundreds place,
9 in the ones place,
4 in the thousands place, and
1 in the ten-thousands place.

Write the number:

___ , ___ ___ ___ , ___ ___ ___

SRB
4

3. a. Measure line segment *PQ* to the nearest inch.

P •———————————————————————————————————• Q

About _____ inches

b. Measure line segment *RS* to the nearest $\frac{1}{2}$ inch.

R •———————————————————• S

About _____ inches

SRB
128

4. Solve mentally or with a paper-and-pencil algorithm.

a.
```
   729
 + 432
 _____
```

b.
```
  9,004
 −  515
 _____
```

SRB
10–15

5. Riley estimated the height of his classroom ceiling. Circle the best estimate.

A 7 m

B 3 m

C 20 m

D 15 m

SRB
130

LESSON 3·6

Flying to London

Suppose you are flying from Charleston, South Carolina, to London, England. You have a connecting flight in Washington, D.C.

1. Your flight to Washington, D.C., leaves Charleston at 7:05 A.M. It lands in Washington at 8:34 A.M. How long was the flight?

    _____

2. Your flight from Washington, D.C., to London is scheduled to leave at 12:02 P.M. The flight time is 7 hours and 19 minutes. At what time does the flight land (in Washington, D.C., time)?

    _____

3. There is a 5-hour time difference between Washington, D.C., and London. What time does your flight land, London time?

    _____

4. Your return flight from London arrives in Washington, D.C., at 4:46 P.M. Your flight to Charleston is scheduled to leave Washington, D.C., 3 hours and 12 minutes later. What time does your flight to Charleston leave?

    _____

LESSON 3•6

Flying to London *continued*

5. Your flight to Charleston is delayed because of stormy weather. It finally leaves at 8:57 P.M. and lands in Charleston 1 hour and 44 minutes later. What time does it land in Charleston?

____

6. A friend who traveled with you lives in New York City. She took a direct flight from London to New York that left at 9:20 A.M., New York time. The flight was 7 hours, 36 minutes long. What time did it land?

____

7. You go to bed at 11:15 P.M. and wake up at 8:37 A.M. How long did you sleep?

____

8. The next day, you spend 45 minutes looking at all the photos you took on your trip. If you finish at 10:25 A.M., what time did you start?

____

LESSON 3·7 Measuring Air Distances

1. Estimate which city listed below is the closest to Washington, D.C. _____

2. Estimate which city is the farthest. _____

3. Measure the shortest distance between Washington, D.C., and each of the cities shown in the table below. Use the globe scale to convert these measurements to approximate air distances.

   a. Record the globe scale. _____ inch → _____ miles

   b. Complete the table.

Distance from Washington, D.C., to	Measurement on Globe (to the nearest $\frac{1}{2}$ inch)	Air Distance (estimated number of miles)
Cairo, Egypt		
Mexico City, Mexico		
Stockholm, Sweden		
Moscow, Russia		
Tokyo, Japan		
Shanghai, China		
Sydney, Australia		
Warsaw, Poland		
Cape Town, South Africa		
Rio de Janeiro, Brazil		
Choose a city. _____		

4. Explain how you used the globe scale to estimate the air distance between Washington, D.C., and Mexico City, Mexico.

_____

_____

_____

_____

LESSON 3·7

Math Boxes

---

1. If 1 centimeter on a map represents 20 kilometers, then

  a. 2 cm represent _____ km.

  b. 5 cm represent _____ km.

  c. 8 cm represent _____ km.

  d. 3.5 cm represent _____ km.

  e. 6.5 cm represent _____ km.

145

---

2. Complete the "What's My Rule?" table and state the rule.

Rule: _____

in	out
40	5
24	3
64	
	4
	9

162–166

---

3. A rock collector has 136 rocks in her collection. She took them to a geologist who said that 57 of them are volcanic. How many of them are not volcanic?

_____

---

4. Solve the riddle. Then use your Geometry Template to draw the shape.

I am a four-sided polygon.

My two short sides are the same length.

My two long sides are the same length.

The sides of the same length are next to each other.

What am I? _____

100

---

5. Make a ballpark estimate. Write a number model to show your strategy.

$2.83 + $0.92 + $3.07 + $7.91

_____

_____

181

---

6. Complete.

  a. 0.1, 0.2, 0.3, _____, _____, _____

  Rule: _____

  b. 1.4, 1.6, 1.8, _____, _____, _____

  Rule: _____

  c. 3, 2.5, 2, _____, _____, _____

  Rule: _____

160 161

LESSON 3·8   Flying to Cairo

Pretend that you are flying from Washington, D.C., to Cairo, Egypt. You have a choice of flying by way of Amsterdam, in the Netherlands, or by way of Rome, Italy. The air distances are shown in the table.

Air Distance between Cities (in miles)

	Amsterdam	Rome
Washington, D.C.	3,851	4,497
Cairo	2,035	1,326

Solve each problem below. Record a number model for the problem using a letter for the unknown. Then write a summary number model with your answer in place of the letter.

1. What is the total distance from Washington, D.C., to Cairo by way of Amsterdam?

   Answer: About _____ miles

   _____
   (number model with unknown)

   _____
   (number model with answer)

2. About how many more miles is it from Washington, D.C., to Rome than from Washington, D.C., to Amsterdam?

   Answer: About _____ miles

   _____
   (number model with unknown)

   _____
   (number model with answer)

3. What is the total distance from Washington, D.C., to Cairo by way of Rome?

Answer: About _____ miles

_____
(number model with unknown)

_____
(number model with answer)

4. About how many fewer miles will you fly if you go by way of Rome rather than Amsterdam?

Answer: About _____ miles

_____
(number model with unknown)

_____
(number model with answer)

Try This

5. If you fly first class, your ticket will cost $9,250. If you fly economy class, you will save $6,500. You bought 2 first class tickets and 4 economy tickets. What is the total cost?

Answer: _____

_____
(number model(s) with unknown)

_____
(number model(s) with answer)

Math Boxes

---

1. Complete the name-collection box.

480

SRB 149

2. A number has

1 in the tens place,
3 in the hundreds place,
5 in the ones place,
7 in the hundred-thousands place,
9 in the thousands place,
2 in the ten-thousands place, and
4 in the millions place.

Write the number:

___ , ___ ___ ___ , ___ ___ ___

SRB 4

---

3. **a.** Measure the line segment to the nearest inch.

B •——————————————————————————• L

About _____ in.

b. Measure the line segment to the nearest $\frac{1}{2}$ inch.

R •————————————————————• T

About _____ in.

SRB 128

---

4. Solve with a paper-and-pencil algorithm.

a.   604
     + 817
     ——

b.   3,005
     − 686
     ——

SRB 10–15

5. Sara estimated the length of her arm. Circle the best estimate.

A   5 centimeters

B   50 centimeters

C   100 centimeters

D   150 centimeters

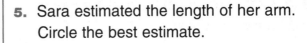
SRB 130

Number Sentences

SRB
148

Tell whether each number sentence below is true or false. Write T for true
or F for false. If it is not possible to tell, write ? on the answer blank.

1. $7 < 3 + 1$ _____   **2.** $6 = 36 \div 6$ _____

3. $80 - ? = 40$ _____   **4.** $28 - 16 = 12$ _____

5. $0 = 4 / 4$ _____   **6.** $2 * 7$ _____

7. $14 \times 3 < 19 \times 2$ _____   **8.** $144 + 76 = 880 \div 4$ _____

9. Make up two true number sentences and two false number sentences.

    a. true _____

    b. true _____

    c. false _____

    d. false _____

10. Make up three true number sentences and three false number sentences. Mix them up.
Ask your partner to write whether each sentence is true or false.

    Example: $4 * 7 = 34 - 6$ T

    a. _____   _____

    b. _____   _____

    c. _____   _____

    d. _____   _____

    e. _____   _____

    f. _____   _____

LESSON 3·9   **Math Boxes**

---

1. If 1 inch on a map represents 30 miles, what would 3 inches represent? Circle the best answer.

   A 10 miles

   B 60 miles

   C 90 miles

   D 300 miles

---

2. Complete the "What's My Rule?" table and state the rule.

Rule: _____

in	out
45	5
81	9
	3
	4
72	

---

3. The Statue of Chief Crazy Horse in South Dakota is 563 feet tall. The Statue of Liberty is 151 feet tall. What is the difference in height of the two statues?

_____ feet

---

4. Solve the riddle. Then use your Geometry Template to trace the shape.

I am a polygon.

All my angles have the same measure.

Each of my 5 sides has the same measure.

What am I?

_____

---

5. Make a ballpark estimate. Write a number model to show your strategy.

$2.50 + $0.75 + $3.85 + $12.70

_____

_____

---

6. Complete.

  a. 5.05, 5.06, 5.07, _____, _____, _____

    Rule: _____

  b. 4, 3.8, 3.6, _____, _____, _____

    Rule: _____

  c. 2.7, 3.2, 3.7, _____, _____, _____

    Rule: _____

LESSON 3·10 Parentheses in Number Sentences

SRB 150

Part 1

Make a true sentence by filling in the missing number.

1. **a.** $(30 - 15) * 2 =$ _____   **b.** $30 - (15 * 2) =$ _____

2. **a.** _____ $= 28 / (14 / 2)$ **b.** _____ $= (28 / 14) / 2$

3. **a.** $(6 + 8) / (2 - 1) =$ _____   **b.** $6 + (8 / 2) - 1 =$ _____

Part 2

Make a true sentence by inserting parentheses.

4. **a.** $4 \times 9 - 2 = 34$ **b.** $4 \times 9 - 2 = 28$

5. **a.** $24 = 53 - 11 + 18$ **b.** $60 = 53 - 11 + 18$

6. **a.** $12 / 4 + 2 = 2$ **b.** $12 / 4 + 2 = 5$

7. **a.** $55 = 15 + 10 \times 4$ **b.** $100 = 15 + 10 \times 4$

Try This

Make a true sentence by inserting two sets of parentheses in each problem.

8. **a.** $10 - 4 / 2 * 3 = 24$ **b.** $10 - 4 / 2 * 3 = 1$

Part 3

Pretend you are playing a game of *Name That Number* with only 3 cards per hand.
To name the target number, use all 3 numbers and any operations you want.
For each problem, write a true number sentence containing parentheses using the
3 numbers and the target number.

9. Use: 2, 5, 15   Target number: 5   _____

10. Use: 3, 4, 5   Target number: 17   _____

11. Use: 1, 3, 11   Target number: 4   _____

LESSON 3·10

Math Boxes

---

1. Complete.

   a. Name all the factors of 50.

      ____ , ____ , ____ , ____ , ____ , ____

   b. Name the factor pairs of 36.

      _____ and _____

      _____ and _____

      _____ and _____

      _____ and _____

      _____ and _____

SRB 7

---

2. In the 2004 Summer Olympics, which two countries had a combined medal count of 155?

_____ and _____

Country	Number of Medals
Australia	49
United States	103
China	63
Russia	92

---

3. What is the mode for the number of books read by the students? Circle the best answer.

Number of Books Read in a Month

  A 3

  B 4

  C 5

  D 2

SRB 73

---

4. Which of these angles has a measure less than 90 degrees? Circle them.

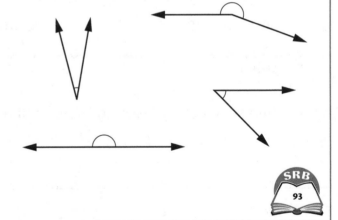

SRB 93

---

5. a. Measure the line segment to the nearest centimeter.

   B ●————————————————————————● *N*

   About _____ cm

  b. Draw a line segment that is half the length of $\overline{BN}$.

  c. How long is the line segment you drew? About _____ cm

SRB 128

LESSON 3·11 Broken Calculator

Solve each open sentence on your calculator without using the "broken" key.
Only one key is broken in each problem. Record your steps.

1.

Broken Key: $\boxed{-}$	
To Solve: $68 + x = 413$	

2.

Broken Key: $\boxed{-}$	
To Solve: $z + 643 = 1{,}210$	

3.

Broken Key: $\boxed{+}$	
To Solve: $d - 574 = 1{,}437$	

4.

Broken Key: $\boxed{\times}$	
To Solve: $w / 15 = 8$	

Try This

5.

Broken Key: $\boxed{\div}$	
To Solve: $s * 48 = 2{,}928$	

6. Make up one for your partner to solve.

Broken Key: $\boxed{}$	
To Solve:	

LESSON 3·11 Open Sentences

Solve each open sentence. Copy the entire sentence with the solution
in place of the variable. Circle the solution.

1. $48 + d = 70$

$48 + (22) = 70$

2. $51 = n + 29$

_____

3. $34 - x = 7$

_____

4. $32 = 76 - p$

_____

5. $h - 6 = 9$

_____

6. $b - 7 = 12$

_____

7. $u - 30 = 10$

_____

8. $5 * m = 35$

_____

9. $y = 3 * 8$

_____

10. $21 / x = 7$

_____

11. $x = 32 / 8$

_____

12. $5 = w / 10$

_____

Try This

13. Mr. O'Connor wrote two open sentences on the board.

$45 + x = 71$
$45 + y = 71$

Isabel says the two open sentences must have different solutions because
the variables are different.

a. Do you agree with Isabel? _____

b. Explain your answer.

_____

_____

_____

LESSON 3·11 Estimating Distances

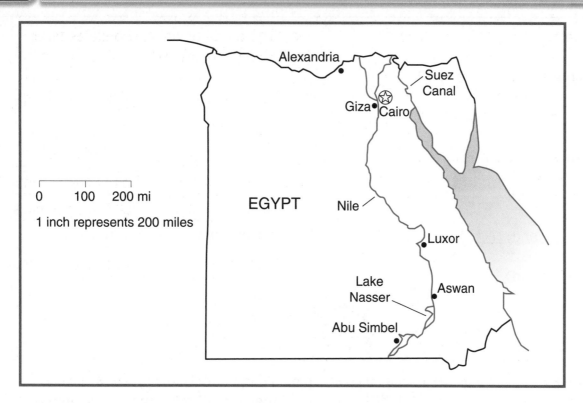

You want to take a trip to Egypt and see the following sights:

◆ Cairo, the capital, on the Nile River, near the Pyramids at Giza

◆ Alexandria, a busy modern city and port on the Mediterranean

◆ The Aswan High Dam across the Nile River, completed in 1970, and Lake Nasser, which formed behind the dam

◆ The temples at Abu Simbel, built more than 3,000 years ago and moved to their present location in the 1960s to escape the rising water of Lake Nasser

You want to know how far it is between locations.

1. The distance between Alexandria and Abu Simbel is about _____ inch(es) on the map.

   That represents about _____ miles.

2. The distance between Cairo and Aswan is about _____ inch(es) on the map.

   That represents about _____ miles.

3. The distance between Abu Simbel and Aswan is about _____ inch(es) on the map.

   That represents about _____ miles.

LESSON 3·11   **Math Boxes**

---

1. Complete.

   a. Name all the factors of 12.

   _____, _____, _____, _____, _____, _____

   b. Name the factor pairs of 16.

   _____ and _____

   _____ and _____

   _____ and _____

SRB 7

---

2. The areas of which two states differ by 944 square miles?

   _____ and _____

State	Total Area
Connecticut	5,543 square miles
Rhode Island	1,545 square miles
Delaware	2,489 square miles
New Jersey	8,721 square miles

---

3. Use the bar graph to answer the questions.

   a. How many students slept 8 hours?

   _____

   b. What is the mode for the number of hours slept?

   _____

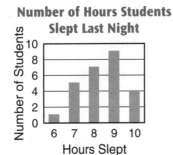

Number of Hours Students Slept Last Night

Number of Students / *Hours Slept*

SRB 73

---

4. Which of the angles below have a measure of more than 90 degrees? Circle them.

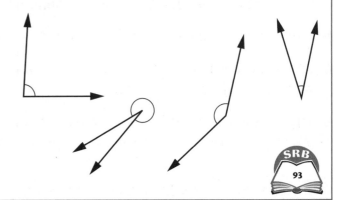

SRB 93

---

5. a. Measure the line segment to the nearest centimeter.

   L •———————————————————————————• P

   About _____ cm

   b. Draw a line segment that is half the length of $\overline{LP}$.

   c. How long is the line segment you drew?   About _____ cm

SRB 128

LESSON 3·12 **Math Boxes**

---

1. Put these numbers in order from smallest to largest.

0.8   0.08   0.73   0.095

_____

_____

_____

_____

SRB
32

2. Complete.

a. 0.1, 0.9, 1.7, _____, _____, _____

Rule: _____

b. 5.6, 5.3, 5, _____, _____, _____

Rule: _____

c. 7.23, 7.28, 7.33, _____, _____, _____

SRB
160 161

Rule: _____

---

3. Solve mentally or with a paper-and-pencil algorithm.

    a. $11.03
         + $4.79

    b. $3.56
         − $2.89

SRB
34–37

4. Blake estimated the length of his little finger. Circle the most reasonable estimate.

A   20 millimeters

B   45 millimeters

C   80 millimeters

D   4 millimeters

SRB
130

---

5. a. Measure the line segment to the nearest centimeter.

A   B

About _____ cm

b. Draw a line segment that is half the length of $\overline{AB}$.

c. How long is the line segment you drew?   About _____ cm

SRB
128

77

LESSON 4·1 Place-Value Number Lines

Fill in the missing numbers.

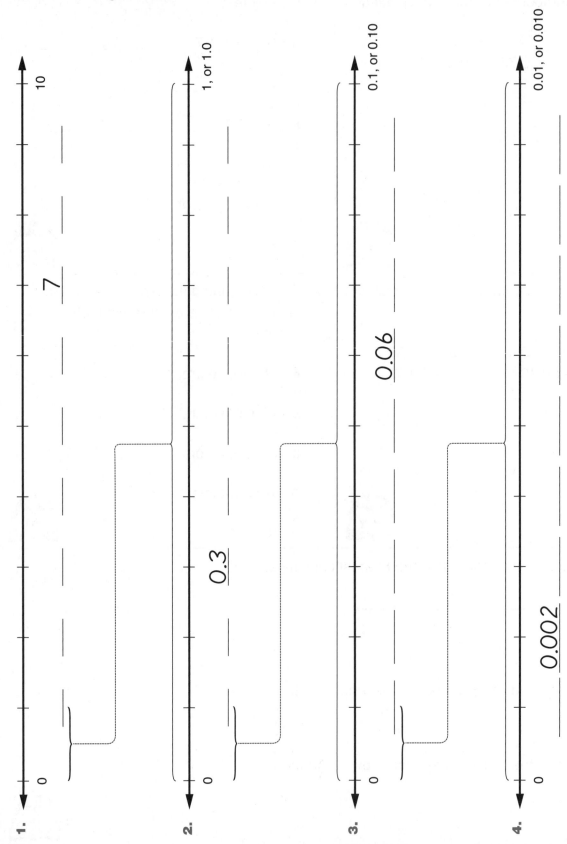

LESSON 4·1 **Math Boxes**

---

1. Solve mentally.

   a. $9 * 4 =$ _____

   b. $6 *$ _____ $= 18$

   c. $3 *$ _____ $= 21$

   d. $16 \div 4 =$ _____

   e. $20 \div 4 =$ _____

   f. $54 \div 6 =$ _____

SRB
16 20

---

2. Solve $199 = p - 408$.
Choose the best answer.

   ⬭ $p = 209$

   ⬭ $p = 309$

   ⬭ $p = 607$

   ⬭ $p = 507$

SRB
148

---

3. In the numeral 9,358,461.72, the 6 is worth 60.

   a. The 4 is worth _____.

   b. The 8 is worth _____.

   c. The 3 is worth _____.

   d. The 9 is worth _____.

   e. The 7 is worth _____.

SRB
30 31

---

4. Draw and label ray *BY*.
Draw point *A* on it.

SRB
91

---

5. Insert parentheses to make these number sentences true.

   a. $5 * 4 - 2 = 18$

   b. $25 + 8 * 7 = 81$

   c. $1 = 36 / 6 - 5$

   d. $19 = 15 - 5 + 81 / 9$

SRB
150

---

6. Estimate the sum. Write a number model to show how you estimated.

$458 + 1,999 + 12,307$

Number model: _____

_____

_____

SRB
181

LESSON 4·2

Tenths and Hundredths

Base-10 Block	Symbol	Value
Flat	□	1
Long	|	$\frac{1}{10}$ or 0.1
Cube	.	$\frac{1}{100}$, or 0.01

1. Complete the table.

Base-10 Blocks	Fraction Notation	Decimal Notation
||	$\frac{2}{10}$	0.2
▪▪▪▪▪.		
||||| |||		
□ ||| ▪▪▪▪▪		

2. Write each number in decimal notation.

Example: $\frac{3}{10}$ = __0.3__

a. $\frac{4}{10}$ = _____

b. $\frac{71}{100}$ = _____

c. $32\frac{6}{100}$ = _____

3. Draw base-10 blocks to show each number. Draw as few blocks as possible.

Example: 0.3   \|\|\|

a. 0.43

b. 2.16

c. 0.07

4. Write each of the following in decimal notation.

a. 8 tenths _____

b. 82 hundredths _____

c. 38 and 5 tenths _____

Try This

5. Write 53 thousandths in decimal notation. _____

80

LESSON 4·2   # Math Boxes

---

1. a. What is the maximum number of blocks any student lives from school?

_____

b. What is the minimum number of blocks?

_____

c. What is the mode? _____

d. What is the median number of blocks?

_____

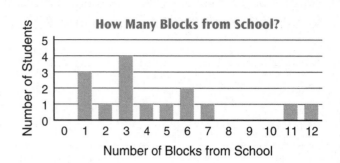

How Many Blocks from School?

Number of Students

Number of Blocks from School

SRB 73

---

2. Solve mentally or with a paper-and-pencil algorithm.

a.   391
       + 467

b.   983
       − 494

SRB 10–15

---

3. If 1 inch on a map represents 40 miles, then

a. 2 in. represent _____ mi.

b. 4 in. represent _____ mi.

c. 5 in. represent _____ mi.

d. $2\frac{1}{2}$ in. represent _____ mi.

e. $1\frac{3}{4}$ in. represent _____ mi.

SRB 145

---

4. Write as dollars and cents.

a. 20 dimes = $_____._____

b. 20 nickels = $_____._____

c. 20 quarters = $_____._____

d. 10 quarters and 7 dimes =

$_____._____

---

5. Solve mentally.

a. 5 * 3 = _____

b. 5 * 30 = _____

c. _____ = 5 * 9

d. _____ = 50 * 9

e. 6 * 7 = _____

f. 60 * 70 = _____

SRB 17

LESSON 4·3   Comparing Decimals

Math Message

1. Arjun thought that 0.3 was less than 0.15. Explain or draw pictures to help Arjun see that 0.3 is more than 0.15.

_____

_____

_____

_____

_____

2. Use base-10 blocks to complete the following table.

> "<" means "is less than."
>
> ">" means "is greater than."

Base-10 Blocks	Decimal	>, <, or =	Decimal	Base-10 Blocks
‖	0.2	>	0.12	❘▪▪
▪▪▪▪▪			0.1	
	0.13			‖‖▪
‖‖ ▪▪▪			0.3	
	1.2			☐☐❘
‖‖‖ ▪▪▪▪▪			0.39	
	2.3			☐‖‖‖‖ ‖‖‖‖ ‖‖‖

Date _____   Time _____

1. Write < or >.

   a. 0.24 ——— 0.18   b. 0.05 ——— 0.1   c. 0.2 ——— 0.35

   d. 1.03 ——— 0.30   e. 3.2 ——— 6.59   f. 25.9 ——— 25.72

2. Write your own decimals to make true number sentences.

   a. _____ > _____   b. _____ < _____   c. _____ < _____

3. Put these numbers in order from smallest to largest.

   a. 0.05, 0.5, 0.55, 5.5

   _____   _____   _____   _____
   smallest   largest

   b. 0.99, 0.27, 1.8, 2.01

   _____   _____   _____   _____
   smallest   largest

   c. 2.1, 2.01, 20.1, 20.01

   _____   _____   _____   _____
   smallest   largest

   d. 0.01, 0.10, 0.11, 0.09

   _____   _____   _____   _____
   smallest   largest

4. Write your own decimals in order from smallest to largest.

   _____   _____   _____   _____
   smallest   largest

5. "What's green inside, white outside, and hops?"
   To find the answer, put the numbers in order from smallest to largest.

0.66	1	0.2	1.05	0.90	0.01	0.75	0.35	$\frac{25}{100}$	$\frac{50}{100}$	0.05	0.09	5.5
N	I	O	C	W	A	D	S	G	A	F	R	H

Write your answers in the following table. The first answer is done for you.

0.01												
A												

LESSON 4·3 **Math Boxes**

---

1. Solve mentally.

    a. $5 * \_\_\_\_\_ = 40$

    b. $9 * 9 = \_\_\_\_\_$

    c. $9 * \_\_\_\_\_ = 27$

    d. $42 \div 6 = \_\_\_\_\_$

    e. $54 \div 9 = \_\_\_\_\_$

    f. $63 \div 7 = \_\_\_\_\_$

SRB 16 20

---

2. Solve each open sentence.

    a. $100 + w = 175$ $w = \_\_\_\_\_$

    b. $503 + y = 642$ $y = \_\_\_\_\_$

    c. $p + 263 = 319$ $p = \_\_\_\_\_$

    d. $444 - s = 93$ $s = \_\_\_\_\_$

    e. $r - 320 = 600$ $r = \_\_\_\_\_$

SRB 148

---

3. In 34.561

    a. The 3 is worth _____.

    b. The 4 is worth _____.

    c. The 5 is worth _____.

    d. The 6 is worth _____.

    e. The 1 is worth _____.

SRB 30 31

---

4. Draw and label ray *CT*.
Draw point *A* on it.

SRB 91

---

5. Insert parentheses to make these number sentences true.

    a. $7 * 8 - 6 = 50$

    b. $13 - 4 * 6 = 54$

    c. $10 = 3 + 49 / 7$

    d. $28 = 28 - 6 + 42 / 7$

SRB 150

---

6. Estimate the sum. Write a number model to show how you estimated.

$3{,}005 + 9{,}865 + 2{,}109$

Number model: _____

_____

_____

SRB 181

LESSON 4·4   A Bicycle Trip

Diego and Alex often take all-day bicycle trips together. During the summer, they took a 3-day bicycle tour. They carried camping gear in their saddlebags for the two nights they would be away from home.

Alex had a **trip meter** that showed miles traveled in tenths of miles. He kept a log of the distances they traveled each day before and after lunch.

Travel Log		
	Distance Traveled	
Timetable	**Before lunch**	**After lunch**
Day 1	27.0 mi	31.3 mi
Day 2	36.6 mi	20.9 mi
Day 3	25.8 mi	27.0 mi

Use estimation to answer the following questions. Do not work the problems out on paper or with a calculator.

1. On which day did they travel the most miles?   _____

2. On which day did they travel the fewest miles?   _____

3. During the whole trip, did they travel more miles before or after lunch? _____

4. Estimate the total distance they traveled. Choose the best answer.

   ⬭ less than 150 miles   ⬭ between 150 and 180 miles

   ⬭ between 180 and 200 miles   ⬭ more than 200 miles

5. Explain how you solved Problem 4.

   _____

   _____

   _____

6. On Day 1, about how many more miles did they travel after lunch than before lunch?

   _____

7. Diego said that they traveled 1.2 more miles before lunch on Day 1 than on Day 3. Alex disagreed. He said they traveled 2.2 more miles. Who is right? Explain your answer.

   _____

   _____

LESSON 4·4

Math Boxes

---

1. a. What is the maximum number of movies a student viewed in a month?

_____

b. What is the minimum number of movies? _____

c. What is the mode? _____

d. What is the median? _____

Number of Movies Viewed in a Month

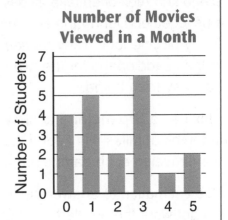

SRB
73

---

2. Solve mentally or with a paper-and-pencil algorithm.

a.   814
      + 123

b.   754
       − 396

SRB
10–15

---

3. If 2 centimeters on a map represent 50 kilometers, then

a. 1 cm represents _____ km.

b. 3 cm represent _____ km.

c. 4 cm represent _____ km.

d. 0.5 cm represent _____ km.

e. 8.5 cm represent _____ km.

SRB
145

---

4. Write 40 quarters and 3 dimes in dollars-and-cents notation. Choose the best answer.

⬭ $4.03

⬭ $4.30

⬭ $10.30

⬭ $40.30

---

5. Solve mentally.

a. 6 * 9 = _____

b. 6 * 90 = _____

c. _____ = 5 * 8

d. _____ = 50 * 8

e. 4 * 4 = _____

f. 4 * 40 = _____

SRB
17

LESSON 4·5 Decimal Addition and Subtraction

Add or subtract mentally or with a paper-and-pencil algorithm.
Pay attention to the + and − symbols.

1. 2.05 + 1.83 = _____

2. 3.04 + 2.8 = _____

3. 2.4 + 3.01 + 0.26 = _____

4. 2.31 − 1.88 = _____

5. 19 + 1.9 = _____

6. 1 − 0.67 = _____

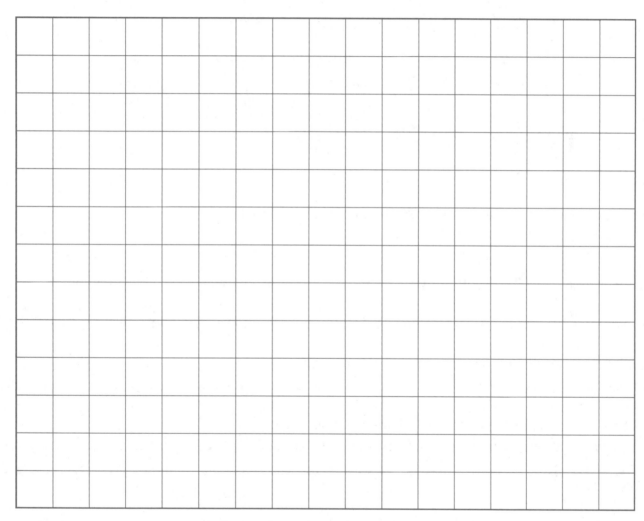

7. Choose one of the problems from above. Explain the method you used
   to solve the problem.

_____

_____

**LESSON
4·5**

Circle Graphs

Percent urban is the number of people out of 100 who live in towns or cities. *Percent rural* is the number of people out of 100 who live in the countryside. Each circle graph below represents the percent of the urban and rural population of an African country.

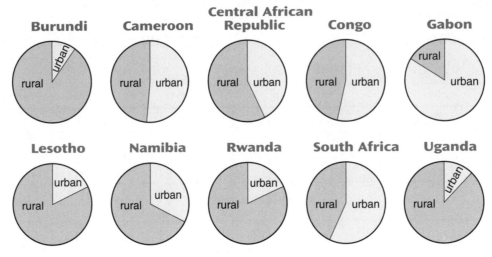

Source: The United Nations

1. For each pair, circle the country with the larger urban population.

    a. Congo   Uganda   b. Rwanda   Gabon

    c. Burundi   South Africa   d. Namibia   Lesotho

2. Which country has the greatest percentage of people living in *urban* areas? _____

3. Which two countries have the greatest percentage
    of people living in *rural* areas?   _____

4. Which two countries have about $\frac{1}{2}$ of their people living
    in urban areas and $\frac{1}{2}$ of their people living in rural areas? _____

Try This

5. Write a question that can be answered from the information in the graphs. Then answer the
    question.

    Question: _____

    _____

    Answer: _____

LESSON 4·5   **Math Boxes**

---

1. Insert >, <, or =.

   a. 0.96 _____ 0.4

   b. 0.50 _____ 0.500

   c. 1.3 _____ 1.09

   d. 0.85 _____ 0.86

   e. 0.700 _____ 0.007

---

2. a. Measure the length of this line segment to the nearest $\frac{1}{2}$ centimeter.

_____

   About _____ cm

  b. Draw a line segment 3 centimeters long.

---

3. Fill in the missing numbers.

   a. 28, 35, 42, _____, _____, _____

   Rule: _____

   b. 56, 48, 40, _____, _____, _____

   Rule: _____

   c. 81, _____, 63, _____, 45, _____

   Rule: _____

---

4. Solve each open sentence.

   a. $5.9 - T = 5$ $T =$ _____

   b. $9.4 - K = 3$ $K =$ _____

   c. $0.81 - M = 0.43$ $M =$ _____

   d. $F - 2.1 = 6.8$ $F =$ _____

   e. $2.43 = S + 1.06$ $S =$ _____

   f. $R - 12.2 = 4.65$ $R =$ _____

---

5. Add 9 tens, 8 hundredths, and 3 tenths to 34.53.

   What is the result? _____

---

6. Add mentally or with a paper-and-pencil algorithm.

   a.
```
        6
       40
      150
  + 1,000
  -------
```

   b.
```
       54
      180
      240
  +   800
  -------
```

SRB
34-37

LESSON 4·6 Keeping a Bank Balance

Math Message

Solve. Show your work on the grid.

1. Cleo went to the store to buy school supplies. She bought a notebook for $2.39, a pen for $0.99, and a set of markers for $3.99. How much money did she spend in all?

   _____

2. Nicholas went to the store with a $20 bill. His groceries cost $13.52. How much change did he get?

   _____

On January 2, Kate's aunt opened a bank account for Kate. Her aunt deposited $100.00 in the account.

Over the next several months, Kate made regular deposits into her account. She deposited part of her allowance and most of the money she made babysitting.

Kate also made a few withdrawals—to buy a radio and some new clothes.

Think about the answers to the following questions:

◆ When you **withdraw** money, do you take money out or put money in?

◆ When you **deposit** money, do you take money out or put money in?

◆ When your money earns **interest,** does this add money to your account or take money away?

The table on the next page shows the transactions (deposits and withdrawals) that Kate made during the first 4 months of the year and the interest she earned.

LESSON 4·6 **Keeping a Bank Balance** *continued*

3. In March, Kate took more money out of her bank account than she put in. In which other month did she withdraw more money than she deposited? _____

4. Estimate whether Kate will have more or less than $100.00 at the end of April. _____

5. Complete the table. Remember to add if Kate makes a deposit or earns interest and to subtract if she makes a withdrawal.

Date	Transaction		Current Balance
January 2	Deposit	$100.00	$ *100.00*
January 14	Deposit	$14.23	+ $ *14.23* $ *114.23*
February 4	Withdrawal	$16.50	$ _____ $ _____
February 11	Deposit	$33.75	$ _____ $ _____
February 14	Withdrawal	$16.50	$ _____ $ _____
March 19	Deposit	$62.00	$ _____ $ _____
March 30	Withdrawal	$104.26	$ _____ $ _____
March 31	Interest	$0.78	$ _____ $ _____
April 1	Deposit	$70.60	$ _____ $ _____
April 3	Withdrawal	$45.52	$ _____ $ _____
April 28	Withdrawal	$27.91	$ _____ $ _____

LESSON 4·6

Math Boxes

1. Solve mentally or with a paper-and-pencil algorithm.

   a. $5.18 − $3.65 = _____

   b. $16.86 + $9.24 = _____

   c. 0.87 + 0.94 = _____

   d. 11.2 − 3.9 = _____

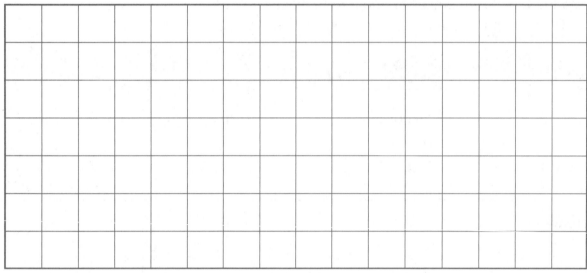

34–37

2. Put these numbers in order from smallest to largest.

   5.92   0.95   9.25   2.95   0.92

   _____ _____ _____ _____ _____

   32 33

3. A trumpeter swan can weigh about 16.8 kilograms. A Manchurian crane can weigh about 14.9 kilograms. How much heavier is a trumpeter swan than a Manchurian crane?

   _____ kilograms

4. Number of items students brought to the school food drive:

   28, 26, 3, 8, 2, 6, 8, 13, 1, 5

   What is the

   a. maximum? _____   **b.** minimum? _____

   c. range? _____   **d.** mode? _____

   e. median? _____   **f.** mean? _____

   73 75

5. How do you write the following number using digits: six-hundred million, five thousand, twenty-one? Choose the best answer.

   ⬭ 6,005,210

   ⬭ 600,500,021

   ⬭ 600,500,210

   ⬭ 600,005,021

   4

LESSON 4·7

Math Boxes

1. Insert >, <, or =.

a. 0.6 _____ 0.57

b. 0.37 _____ 0.36

c. 2.56 _____ 2.056

d. 0.24 _____ 0.240

e. 0.008 _____ 0.080

32 33

2. a. Measure the length of this line segment to the nearest $\frac{1}{2}$ centimeter.

_____

About _____ cm

b. Draw a line segment 7 centimeters long.

128

3. Fill in the missing numbers.

a. 16, 20, 24, _____, _____, _____

Rule: _____

b. 15, _____, 21, _____, 27, _____

Rule: _____

c. _____, _____, 30, _____, _____, 60

Rule: _____

160 161

4. Solve each open sentence.

a. $D + 1.0 = 9.2$ $D =$ _____

b. $5.6 + K = 10$ $K =$ _____

c. $0.8 - M = 0.6$ $M =$ _____

d. $9.09 + S = 13.64$ $S =$ _____

e. $F + 1.25 = 12.90$ $F =$ _____

f. $R - 0.03 = 1.65$ $R =$ _____

148

5. Add 6 tens, 4 hundredths, and 2 tenths to 367.53.

What is the result? _____

36

6. Add mentally or with a paper-and-pencil algorithm.

a.
```
      24
      80
     360
 + 1,200
```

b.
```
       6
     180
      40
 + 1,200
```

10 11

93

LESSON 4·7 Tenths, Hundredths, and Thousandths

Math Message

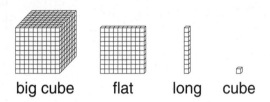

big cube   flat   long   cube

1. 1 big cube = _____ flats

   1 flat = $\dfrac{1}{\boxed{10}}$ of a big cube

2. 1 big cube = _____ longs

   1 long = $\dfrac{1}{\boxed{}}$ of a big cube

3. 1 big cube = _____ cubes

   1 cube = $\dfrac{1}{\boxed{}}$ of a big cube

Base-10 Block	Symbol	Value
Big Cube	⬜	1
Flat	☐	$\frac{1}{10}$, or 0.1
Long	|	$\frac{1}{100}$, or 0.01
Cube	▪	$\frac{1}{1,000}$, or 0.001

4. Complete the table.

Base-10 Blocks	Fraction Notation	Decimal Notation
☐ ||| ..	$\frac{142}{1,000}$	0.142
☐ ☐		
|||| |		
☐ ☐ ☐ ☐ ☐		
☐ ||| ...		

94

LESSON 4·7 **Tenths, Hundredths, and Thousandths** *cont.*

5. Write each number in decimal notation.

Example: $\frac{72}{1,000}$ = _0.072_

a. $\frac{416}{1,000}$ = _____

b. $\frac{17}{1,000}$ = _____

c. $\frac{8}{1,000}$ = _____

d. $7\frac{9}{100}$ = _____

e. $8\frac{14}{1,000}$ = _____

f. $385\frac{4}{10}$ = _____

6. Draw base-10 blocks to show each number.

Example: 0.205 ☐ ☐ ▸ ▴ ▾ ▴ ▾

a. 0.21

b. 0.306

c. 0.008

d. 1.054

7. Write each of the following in decimal notation.

a. 284 thousandths _____

b. 36 hundredths _____

c. 7 thousandths _____

d. 90 and 16 thousandths _____

e. 15 and 3 tenths _____

f. 408 thousandths _____

8. Write < or >.

a. 0.302 _____ 0.203

b. 0.51 _____ 0.310

c. 0.816 _____ 0.9

d. 1.47 _____ 1.5

e. 0.073 _____ 0.73

f. 4.01 _____ 4.009

Try This

9. Put these numbers in order from smallest to largest: 0.03, 0.009, 0.285, 0.064

_____ _____ _____ _____
smallest   largest

95

| LESSON 4·8 | **Measuring Length with Metric Units** |

1. Your teacher will choose several objects or distances to measure. Measure each to the nearest centimeter. Then compare your measurements with your partner's. If you do not agree, work together to measure the objects again. Record the results in the table.

SRB
128

Object or Distance	My Measurement	Partner's Measurement	Agreed Measurement
	About _____ cm	About _____ cm	About _____ cm
	About _____ cm	About _____ cm	About _____ cm
	About _____ cm	About _____ cm	About _____ cm
	About _____ cm	About _____ cm	About _____ cm
	About _____ cm	About _____ cm	About _____ cm

2. Measure these line segments to the nearest centimeter.

   a. _____

   About _____ centimeters

   b. _____

   About _____ centimeters

3. Measure these line segments to the nearest $\frac{1}{2}$ centimeter.

   a. _____

   About _____ centimeters

   b. _____

   About _____ centimeters

LESSON 4·8 **Math Boxes**

---

1. Solve mentally or with a paper-and-pencil algorithm.

a. 3,309
   + 721

b. 2,700
   − 1,299

SRB
10–15

---

2. Complete.

a. 1 cm = _____ mm

b. 5 cm = _____ mm

c. _____ cm = 30 mm

d. 100 cm = _____ mm

e. 200 cm = _____ mm

SRB
129

---

3. Tell whether each number sentence is true or false.

a. $8.77 - 0.08 = 8.50$ _____

b. $35.7 + 22.1 = 57.87$ _____

c. $90.2 - 44.9 < 45$ _____

d. $4.66 + 2.13 > 6$ _____

SRB
36 37
148

---

4. Trace at least two regular polygons from your Geometry Template.

SRB
97

---

5. Without measuring, estimate the length of your foot from heel to toe. Then measure the length of your foot.

a. Estimate:

About _____ cm

b. Measurement:

About _____ cm

SRB
128 130

---

6. Complete.

a. Is 47 closer to 40 or 50?

_____

b. Name the number halfway between 30 and 40.

_____

SRB
182 183

LESSON 4·9   Personal References for Units of Length

Personal References for Metric Units of Length

Use a ruler, meterstick, or tape measure to find common objects that have lengths of 1 centimeter, 1 decimeter, and 1 meter. The lengths do not have to be exact, but they should be close. Ask a friend to look for references with you. You can find more than one reference for each unit. Record the references in the table below.

Unit of Measure	Personal References
1 centimeter (cm)	
1 decimeter (dm), or 10 centimeters	
1 meter (m)	

To be completed in Lesson 5-1.

Personal References for U.S. Customary Units of Length

Use a ruler, yardstick, or tape measure to find common objects that have lengths of 1 inch, 1 foot, and 1 yard. The lengths do not have to be exact, but they should be close. Ask a friend to look for references with you. You can find more than one reference for each unit. Record the references in the table below.

Unit of Measure	Personal References
1 inch (in.)	
1 foot (ft)	
1 yard (yd)	

LESSON 4·9 Measurement Collection for Metric Units of Length

Use your personal references to estimate the length of an object or a distance in centimeters, decimeters, or meters. Describe the object or distance and record your estimate in the table below. Then measure the object or distance and record the actual measurement in the table.

SRB
130

Object or Distance	Estimated Length	Actual Length

**LESSON
4·9**   **Math Boxes**

---

1. Solve mentally or with a paper-and-pencil algorithm.

   a. $12.63 + $5.66 = _____

   b. $2.46 − $1.34 = _____

   c. 9.6 − 4.8 = _____

   d. 0.64 + 0.47 = _____

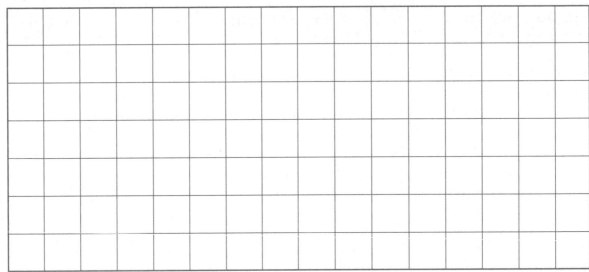

34–37

---

2. Put these numbers in order from smallest to largest.

   1.68   0.78   6.71   0.61

   _____ _____ _____ _____
   32 33

3. The great spotted kiwi bird is about 114.3 centimeters tall. The greater rhea is about 137.1 centimeters tall. How much taller is the greater rhea than the great spotted kiwi bird?

   _____ centimeters

---

4. Make up a set of 7 numbers having the following landmarks:

   mode: 21
   median: 24
   maximum: 35
   range: 20

   ___, ___, ___, ___, ___, ___, ___
   73

5. Write the following numbers using digits:

   a. four hundred eighty-two thousand, one hundred ninety-seven

      _____

   b. eight hundred million, twelve thousand, five

      _____

   4

LESSON
4·10 **Measuring in Millimeters**

Math Message

On your centimeter ruler, the numbered marks are for centimeters and
the little marks between the centimeter marks are for millimeters.

1. Look at your centimeter ruler. How many millimeters are in 1 centimeter? _____ mm

2. Name something that measures about 1 millimeter. _____

    _____

    _____

3. Draw a line segment that is 8 centimeters long.

4. Draw a line segment that is 80 millimeters long.

Measure each line segment below using both the millimeter side and the
centimeter side of the cm/mm ruler. Record both measurements.

5. *A* _____ *B*

    Length of $\overline{AB}$ = _____ mm = _____ cm

6. *C* _____ *D*   Length of $\overline{CD}$ = _____ mm = _____ cm

7. *E* *F*   Length of $\overline{EF}$ = _____ mm = _____ cm

Measuring Land Invertebrates

An invertebrate is an animal that does not have a backbone. (The backbone is also
called the spinal column.) Some invertebrates live on land, others in water. The most
common land invertebrates are insects.

The invertebrates shown on page 102, except the earthworm, bumblebee, and
mealybug, have been drawn to about actual size. The earthworm can grow to about
4 times the length shown. The bumblebee is shown about twice its actual size and the
mealybug about 3 times its actual size.

101

Measuring Land Invertebrates

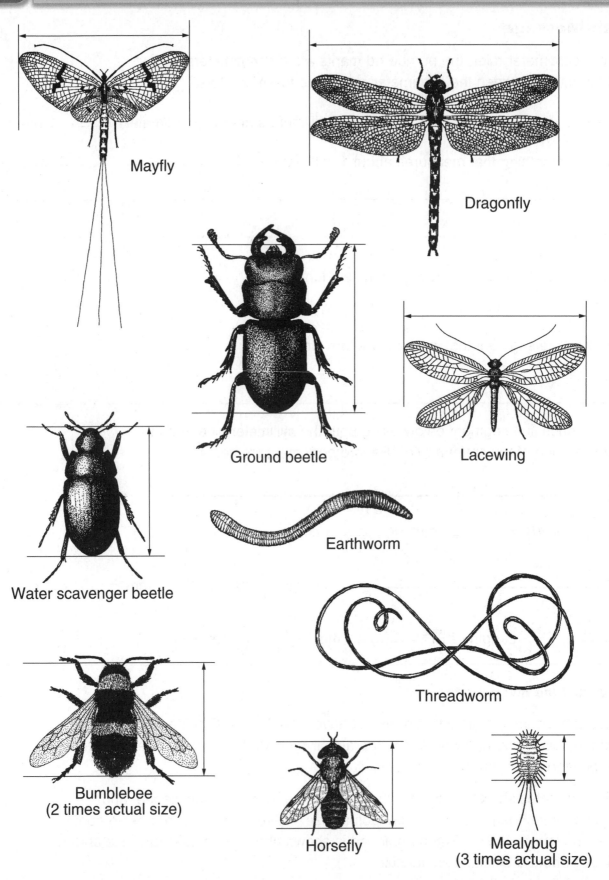

Mayfly

Dragonfly

Ground beetle

Lacewing

Water scavenger beetle

Earthworm

Threadworm

Bumblebee
(2 times actual size)

Horsefly

Mealybug
(3 times actual size)

LESSON 4·10   Measuring Land Invertebrates *continued*

Refer to the pictures on page 102 to answer the following questions.

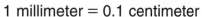

| 1 centimeter (cm) = 10 millimeters (mm) |
| 1 millimeter = 0.1 centimeter |

SRB
128 129

1. Measure the following invertebrates to the nearest millimeter by finding the distance between the two guidelines. Then give the lengths in centimeters.

   a. mayfly   About _____ mm   About _____ cm

   b. dragonfly   About _____ mm   About _____ cm

   c. water scavenger beetle   About _____ mm   About _____ cm

   d. ground beetle   About _____ mm   About _____ cm

   e. lacewing   About _____ mm   About _____ cm

   f. horsefly   About _____ mm   About _____ cm

2. How much longer is the ground beetle than the water scavenger beetle? About _____ cm

3. The bee has been drawn to twice its actual size.
   In reality, which is longer, the bee or the horsefly? _____

   How much longer? About _____ mm

4. The mealybug has been drawn to 3 times its actual size. In the space at the right, draw a mealybug that is about the actual size.

5. What is the actual size of the mealybug in millimeters? _____ mm

6. How did you solve Problem 5?

   _____

   _____

   _____

7. When straight, the threadworm in the drawing is 306 millimeters long.

   What is its length in centimeters? _____ cm   In meters? _____ m

LESSON 4·10   **Math Boxes**

1. Solve mentally or with a paper-and-pencil algorithm.

   a.   4,647
          + 3,228

   b.   2,500
          − 1,398

SRB
10–15

2. Complete.

   a. 7 cm = _____ mm

   b. 15 cm = _____ mm

   c. 500 cm = _____ m

   d. _____ cm = 40 mm

   e. _____ cm = 8 m

SRB
129

3. Tell whether each number sentence is true or false.

   a. $2.34 - 0.09 = 2.25$ _____

   b. $89.6 + 21.7 = 111.3$ _____

   c. $56.4 - 23.8 < 33$ _____

   d. $5.17 + 3.86 > 10$ _____

SRB
36 37
148

4. Name two properties of a regular polygon.

   a. _____

       _____

   b. _____

       _____

SRB
97

5. Without measuring, estimate the height of your chair. Then measure it.

   a. Estimate:

       About _____ cm

   b. Measurement:

       About _____ cm

SRB
128 130

6. Complete.

   a. Is 326 closer to 300 or 400?

       _____

   b. Name the number halfway between 500 and 800.

       _____

SRB
182 183

LESSON
4·11

Math Boxes

---

1. Estimate the sum. Write a number model to show how you estimated.

   3,721 + 2,876 + 7,103

   Number model: _____

   _____

   _____

   SRB
   181

---

2. Solve mentally.

   a. 4 * 8 = _____

   b. 4 * 80 = _____

   c. _____ = 5 * 3

   d. _____ = 50 * 3

   e. 6 * 6 = _____

   f. 6 * 60 = _____

   SRB
   16 17

---

3. Complete.

   a. Is 63 closer to 60 or 70? _____

   b. What number is halfway between 80 and 90? _____

   c. Is 572 closer to 500 or 600? _____

   d. What number is halfway between 300 and 600? _____

   SRB
   182 183

---

4. Write the following numbers using digits:

   a. one million, three hundred forty-six thousand, thirteen

      _____

   b. twenty-two million, fifteen thousand, three hundred fifty-four

      _____

   SRB
   4

---

5. Add mentally or with a paper-and-pencil algorithm.

a.	b.	c.	d.
35	18	54	48
100	420	180	720
280	120	360	180
+ 800	+ 2,800	+ 1,200	+ 2,700

   2 + 2 + 2 + 2

   SRB
   10 11

LESSON 5·1

Multiplying Ones by Tens

You can extend a multiplication fact by making one of the factors a multiple of ten.

Example:

Original fact: 2 * 3 = 6

Extended facts: 2 * **30** = _____, or **20** * 3 = _____

Write a multiplication fact for each Fact Triangle shown below.
Then extend this fact by changing one factor to a multiple of ten.

1.

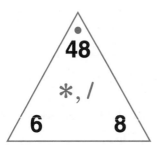

Original fact: _____

Extended fact: _____

2.

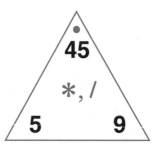

Original fact: _____

Extended fact: _____

3.

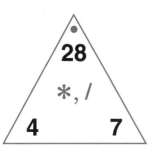

Original fact: _____

Extended fact: _____

4.

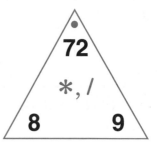

Original fact: _____

Extended fact: _____

5. What shortcut can you use to multiply ones by tens, such as 3 * 60?

_____

_____

LESSON 5·1 **Multiplying Tens by Tens**

You can extend a multiplication fact by making both factors multiples of ten.

Example:

Original fact: 3 * 5 = 15

Extended fact: **30** * **50** = —————

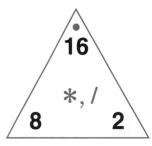

Write a multiplication fact for each Fact Triangle shown below.
Then extend this fact by changing both factors to multiples of ten.

1.

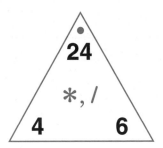

Original fact: _____

Extended fact: _____

2.

Original fact: _____

Extended fact: _____

3.

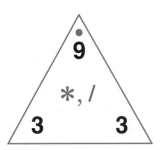

Original fact: _____

Extended fact: _____

4.

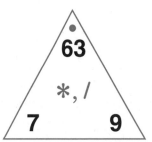

Original fact: _____

Extended fact: _____

5. What shortcut can you use to multiply tens by tens, such as 40 * 60?

_____

_____

LESSON 5·1

Math Boxes

---

1. A number has

6 in the tenths place,
9 in the hundreds place,
2 in the thousands place,
7 in the ones place,
3 in the tens place, and
5 in the hundredths place.

Write the number.

_____ , _____ _____ _____ . _____ _____

31

---

2. Solve mentally.

a. 40 * 50 = _____

b. 70 * 300 = _____

c. 60 * _____ = 180

d. 90 * _____ = 810

e. _____ * 9 = 7,200

17

---

3. Solve mentally or with a paper-and-pencil algorithm.

a.	4,500	**b.**	2,100
	540		420
	100		90
+	12	+	18

10 11

---

4. a. List all the factors of 28.

_____

_____

b. Which of these factors are prime?

_____

7 8

---

5. If 1 inch on a map represents 300 miles, then

a. 6 in. → _____ miles

b. 10 in. → _____ miles

c. _____ in. → 900 miles

d. _____ in. → 750 miles

e. $8\frac{1}{2}$ in. → _____ miles

145

---

6. a. Five children share 27 tennis balls equally.

Each child gets _____ balls.

There are _____ balls left over.

b. There are 32 cookies for 6 friends.

Each friend gets _____ cookies.

There are _____ cookies left over.

20

LESSON 5·2 Presidential Information

The following table shows the dates on which the most recent presidents of the United States were sworn in and their ages at the time they were sworn in.

President	Date Sworn In	Age
F. D. Roosevelt	March 4, 1933	51
Truman	April 12, 1945	60
Eisenhower	January 20, 1953	62
Kennedy	January 20, 1961	43
Johnson	November 22, 1963	55
Nixon	January 20, 1969	56
Ford	August 9, 1974	61
Carter	January 20, 1977	52
Reagan	January 20, 1981	69
G. H. Bush	January 20, 1989	64
Clinton	January 20, 1993	46
G. W. Bush	January 20, 2001	54

1. What is the median age (the middle age) of the presidents at the time they were sworn in? _____ years

2. What is the range of their ages (the difference between the ages of the oldest and the youngest)? _____ years

3. Who was president for the longest time? _____

4. Who was president for the shortest time? _____

5. Presidents are elected to serve for 1 term. A term lasts 4 years. Which presidents served only 1 term or less than 1 term?

_____

6. Which president was sworn in about 28 years after Roosevelt? _____

7. Roosevelt was born on January 30, 1882. If he were alive today, about how old would he be? _____ years old

LESSON 5·2 **Math Boxes**

---

1. Fill in the missing numbers on each number line.

a.
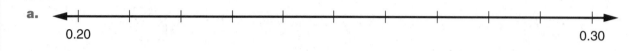

0.20   0.30

___ ___ ___ ___ ___ ___ ___ ___ ___

b.

2.6   2.67   2.7

___ ___ ___ ___ ___ ___ ___ ___   ___ ___

---

2. Measure the line segment to the nearest millimeter. Record the measurement in millimeters and centimeters.

_____

a. About _____ mm

b. About _____ cm _____ mm

c. About _____._____ cm

SRB
128 129

---

3. Tell whether each number sentence is true or false. Write T for true or F for false.

a. $7.89 - 3.36 = 4.53$ _____

b. $12.6 - 4.8 = 3.9 + 3.9$ _____

c. $(4.6 + 2.9) - 3.1 < 3.7$ _____

d. $0.20 > 0.68 - (0.42 + 0.11)$ _____

SRB
34–37
148

---

4. Insert > or < to make a true number sentence.

a. 4,500,999 _____ 879,662

b. 23,468,000 _____ 23,467,000

c. 568,009,352 _____ 568,010,320

d. 400,632 _____ 399,800

SRB
6

---

5. Divide mentally.

a. $45 / 9 =$ _____

b. $450 / 90 =$ _____

c. $1,000 / 200 =$ _____

d. _____ $= 2,000 / 40$

e. _____ $= 6,300 / 700$

SRB
21

Math Boxes

1. A number has

   2 in the hundreds place,
   7 in the tenths place,
   6 in the hundredths place,
   4 in the ones place,
   5 in the tens place, and
   1 in the thousandths place.

Write the number.

____ ____ ____ . ____ ____ ____

31

2. Solve mentally.

a. $3 * 40 =$ _____

b. $90 * 70 =$ _____

c. $50 *$ _____ $= 3,000$

d. _____ $* 8 = 4,000$

e. $80 *$ _____ $= 56,000$

17

3. Solve mentally or with a paper-and-pencil algorithm.

a.
     72
   450
   160
+ 1,000

b.
     15
   240
   350
+ 5,600

10 11

4. a. List all the factors of 50.

_____

_____

b. Which of these factors are prime?

_____

7 8

5. If 1 centimeter on a map represents 200 miles, what do 4.5 centimeters represent? Fill in the circle next to the best answer.

○ **A.** 850 miles

○ **B.** 900 miles

○ **C.** 450 miles

○ **D.** 800 miles

43

6. a. Sara collected 30 leaves. On the way to school, she lost 2 of them. At school she and her 6 friends shared them equally. How many leaves did each person get?

_____ leaves

b. Ava and her 3 sisters shared 24 mints equally. How many mints did each sister get?

_____ mints

20

LESSON 5·3 **Estimated U.S. Distances and Driving Times**

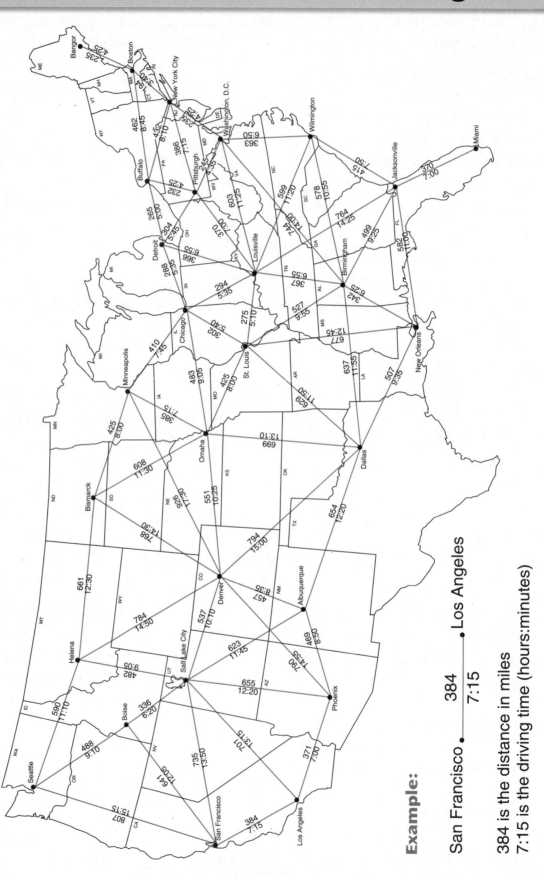

Example:

San Francisco •————384————• Los Angeles
                    7:15

384 is the distance in miles

7:15 is the driving time (hours:minutes)

LESSON 5·3 Planning a Driving Trip

Use the map on journal page 112. Start at your hometown. Plan a driving trip that takes you to 4 other cities on the map. If your hometown is not on the map, find the nearest city on the map to your hometown. Start your driving trip from this city.

Example: Start in Chicago. Drive to St. Louis, Louisville, Birmingham, and then New Orleans.

1. Record your routes, driving distances, and driving times in the table.

From...To	Driving Distance (miles)	Driving Time (hours:minutes)	Rounded Time (hours)
	_____ _____ _____	_____ : _____	_____
	_____ _____ _____	_____ : _____	_____
	_____ _____ _____	_____ : _____	_____
	_____ _____ _____	_____ : _____	_____

2. Estimate how many *miles* you will drive in all.   About _____ miles

3. Estimate how many *hours* you will drive in all.   About _____ hours

4. Tell how many *days* it will take to complete the trip if you plan to drive about 8 hours each day and then stop somewhere for the night.   _____ days

5. Explain how you solved Problems 2–4.

_____

_____

_____

_____

LESSON 5·4   What Do Americans Eat?

The U.S. Department of Agriculture conducts a survey to find out how much food Americans eat. A large number of people are asked to keep lists of all the foods they eat over several days.

These lists are then used to estimate how much of each food was eaten during one year. The average American eats more than 2,000 pounds of food per year. This is about $5\frac{1}{2}$ pounds of food per day.

Results show that the average American eats or drinks about the following amounts in one year:

16	pounds of apples
27	pounds of bananas
5	pounds of broccoli
10	pounds of carrots
30	pounds of cheese
255	eggs
16	pounds of fish
28	cups of yogurt
17	pounds of ice cream
350	cups of milk
22	pounds of candy

Use your answers to the Math Message question to complete these statements.

1. I will eat about _____ eggs in one year.

2. I will drink about _____ cups of milk in one year.

3. I will eat about _____ cups of yogurt in one year.

4. Based on your answers to Problems 1–3, do you think you eat like an average American? Explain why or why not.

_____

_____

LESSON 5·4 Estimating Averages

◆ Estimate whether the answer will be in the tens, hundreds, thousands, or more.

◆ Write a number model to show how you estimated.

◆ Then circle the box that shows your estimate.

Example: Alice sleeps an average of 9 hours per night. How many hours does she sleep in 1 year?

Number model: $10 * 400 = 4,000$

10s	100s	⟨1,000s⟩	10,000s	100,000s	1,000,000s

1. An average of about 23 new species of insects are discovered each day. About how many new species are discovered in one year?

Number model: _____

10s	100s	1,000s	10,000s	100,000s	1,000,000s

2. A housefly beats its wings about 190 times per second. That's about how many times per minute?

Number model: _____

10s	100s	1,000s	10,000s	100,000s	1,000,000s

3. A blue whale weighs about as much as 425,000 kittens. About how many kittens weigh as much as 4 blue whales?

Number model: _____

10s	100s	1,000s	10,000s	100,000s	1,000,000s

4. An average bee can lift about 300 times its own weight. If a 170-pound person were as strong as a bee, about how many pounds could this person lift?

Number model: _____

10s	100s	1,000s	10,000s	100,000s	1,000,000s

LESSON 5·4

Math Boxes

1. Fill in the missing numbers on each number line.

   a.

   5.0   6.0

   ___ ___ ___   ___ ___ ___ ___

   b.

   4.73   5.43   5.73

   ___ ___ ___ ___ ___ ___   ___ ___

2. Measure the line segment to the nearest millimeter. Record the measurement in millimeters and centimeters.

   _____

   a. About _____ mm

   b. About _____ cm _____ mm

   c. About _____ . _____ cm

   128   129

3. Which number sentence is true? Fill in the circle next to the best answer.

   ○ **A.** $(7.6 + 3.8) - 5.2 < 5.9$

   ○ **B.** $0.50 < 1.43 - (0.77 + 0.19)$

   ○ **C.** $14.46 - 12.09 = 3.46$

   ○ **D.** $15.8 - 11.5 = 1.6 + 2.7$

   34–37
   148

4. Insert > or < to make a true number sentence.

   a. 9,000,000 _____ 8,999,999

   b. 421,936,500 _____ 422,985,600

   c. 68,004,002 _____ 68,004,100

   d. 600,523 _____ 601,000

   6

5. Divide mentally.

   a. $18,000 / 9 =$ _____

   b. $350 / 7 =$ _____

   c. $3,500 / 5 =$ _____

   d. _____ $= 5,600 / 800$

   e. _____ $= 42,000 / 60$

   21

Math Boxes

---

1. In 2002, about 32,500,000 people living in the United States had been born in other countries. The circle graph shows where these people were born.

a. Where were most of these people born?

_____

b. About what fraction of the people

were born in Asia? _____

Immigration Statistics

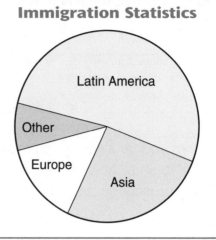

---

2. Estimate the sum. Write a number model to show how you estimated.

a. 387 + 945 + 1,024

Number model:

_____

b. 582 + 1,791 + 2,442

Number model:

_____

181

---

3. Complete.

Rule: _____

in	out
4	280
80	
	490
90	6,300
300	

162–166

---

4. Write two hundred million, three thousand, eighty-eight using digits. Fill in the circle next to the best answer.

○ **A.** 2,030,088

○ **B.** 200,030,088

○ **C.** 20,003,880

○ **D.** 200,003,088

4

---

5. Look at the grid below.

a. In which column is the circle located?

_____

b. In which row is the circle located?

_____

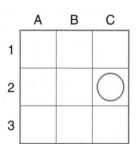

144

LESSON 5·5

The Partial-Products Method

SRB
18

Multiply using the partial-products method. Show your work in the computation grid on page 119.

1. 82 * 3 = _____

2. 6 * 53 = _____

3. 574 * 5 = _____

4. 3 * 470 = _____

5. 2 * 1,523 = _____

6. 3,467 * 3 = _____

Estimate whether your answer will be in the tens, hundreds, thousands, or more. Write a number model to show how you estimated. Circle the correct box. Then calculate the answer. Show your work on page 119.

7. China has the world's longest school year at 251 days. How many school days are in 7 years?

a. Number model: _____

10s	100s	1,000s	10,000s	100,000s	1,000,000s

b. Calculate the answer. _____ days of school

8. People living in the United States eat about 126 pounds of fresh fruit in one year. About how many pounds of fresh fruit would a family of 6 eat in one year?

a. Number model: _____

10s	100s	1,000s	10,000s	100,000s	1,000,000s

b. Calculate the answer. About _____ pounds of fresh fruit

9. Explain how estimation can help you decide whether an answer to a multiplication problem makes sense.

_____

_____

_____

LESSON 5·5

The Partial-Products Method *continued*

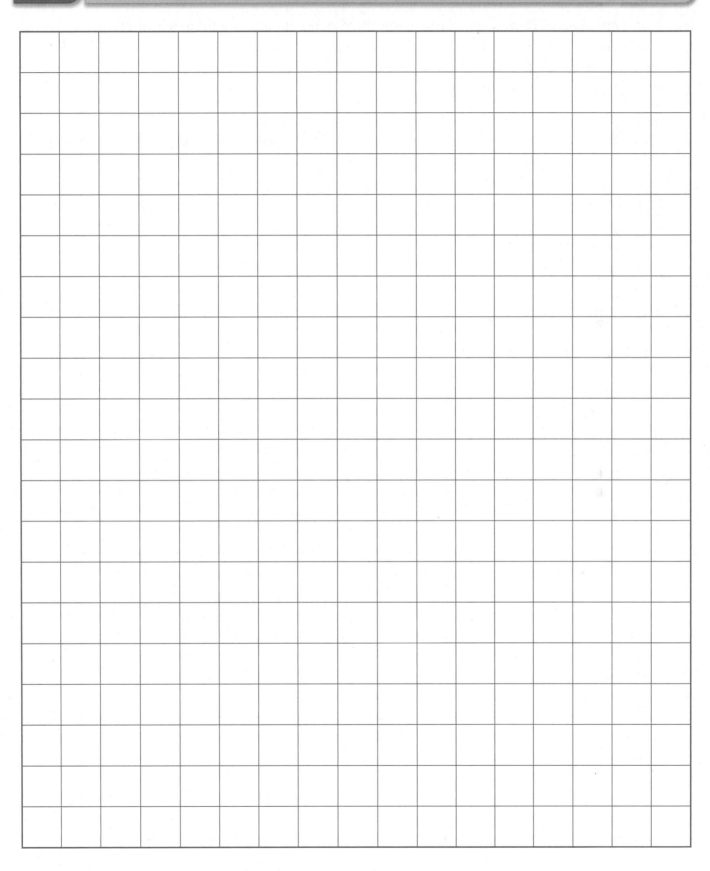

My Measurement Collection for Units of Length

Use your personal references on journal page 98 to estimate the length or height of an object or distance in inches, feet, or yards. Describe the object or distance, and record your estimate in the table below. Then measure the object or distance, and record the actual measurement in the table.

Object or Distance	Estimated Length	Actual Length

LESSON 5·6

Math Boxes

1. **a.** Measure the line segment to the nearest $\frac{1}{4}$ inch.

_____

About _____ inches

b. Draw a line segment that is half as long as the one above.

c. How long is the line segment you drew? About _____ inches

 SRB 128

2. Estimate the product. Write a number model to show how you estimated.

a. 48 * 21

Number model:

_____

b. 98 * 72

Number model:

_____

SRB 184

3. Multiply. Use the partial-products method.

_____ = 52 * 43

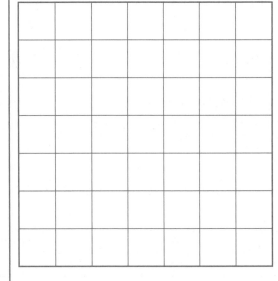

SRB 18

4. Write each number using digits.

a. three hundred forty-two thousandths

_____

b. six and twenty-five hundredths

_____

SRB 27 28

5. If you remove 7 gallons per day from a 65-gallon water tank, how many days will it take to empty the tank?

_____

 SRB 175

Date _____ Time _____

LESSON 5·6 Multiplication Number Stories

Follow these steps for each problem.

a. Decide which two numbers need to be multiplied to give the exact answer.
Write the two numbers.

b. Estimate whether the answer will be in the tens, hundreds, thousands, or more.
Write a number model for the estimate. Circle the box to show your estimate.

c. On the grid below, find the exact answer by multiplying the two numbers.
Write the answer.

1. The average person in the United States drinks about 61 cups of soda per month.
About how many cups of soda is that per year?

a. _____ * _____   b. _____   c. _____

   numbers that give   number model for your estimate   exact answer
   the exact answer

10s	100s	1,000s	10,000s	100,000s	1,000,000s

2. Eighteen newborn hummingbirds weigh about 1 ounce. About how many of them
does it take to make 1 pound? (1 pound = 16 ounces)

a. _____ * _____   b. _____   c. _____

   numbers that give   number model for your estimate   exact answer
   the exact answer

10s	100s	1,000s	10,000s	100,000s	1,000,000s

LESSON 5·6 **Multiplication Number Stories** *continued*

3. A test found that a lightbulb lasts an average of 63 days after being turned on. About how many hours is that?

a. _____ * _____   **b.** _____   **c.** _____

numbers that give
the exact answer

number model for your estimate

exact answer

10s	100s	1,000s	10,000s	100,000s	1,000,000s

4. A full-grown oak tree loses about 78 gallons of water through its leaves per day. About how many gallons of water is that per year?

a. _____ * _____   **b.** _____   **c.** _____

numbers that give
the exact answer

number model for your estimate

exact answer

10s	100s	1,000s	10,000s	100,000s	1,000,000s

123

LESSON 5·7 **Lattice Multiplication**

Use the lattice method to find the products.

1. 3 * 56 = _____

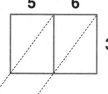

2. 8 * 26 = _____

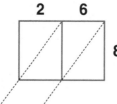

3. 7 * 74 = _____

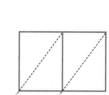

4. 6 * 315 = _____

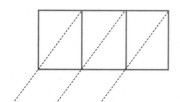

5. 9 * 284 = _____

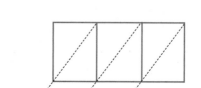

6. 47 * 63 = _____

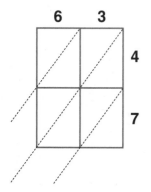

7. 26 * 26 = _____

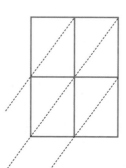

8. 16 * 473 = _____

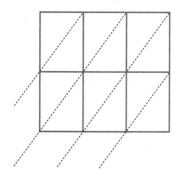

9. 46 * 805 = _____

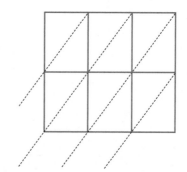

**LESSON
5·7**

Math Boxes

---

1. This is a circle graph of what Seema does on a typical day.

Seema's 24-Hour Day

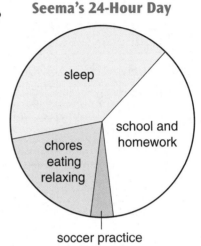

a. What does she spend the least amount of time doing?

_____

b. She spends about the same amount of time at school and doing homework as she does

_____.

c. About what fraction of the day does she spend doing chores, eating, and relaxing? _____

---

2. Estimate the sum. Write a number model to show how you estimated.

a. 715 + 1,904 + 688

Number model:

_____

b. 867 + 2,346 + 3,596

Number model:

_____

SRB
181

---

3. Complete.

Rule: _____

in	out
50	4,000
70	
	7,200
100	
45	3,600

SRB
162–166

---

4. Write each number using digits.

a. twenty-six million, nineteen thousand, eighteen

_____

b. three hundred fifty-two million, eight hundred thousand, two hundred

_____

SRB
4

---

5. Look at the grid below.

a. In which column is the triangle located?

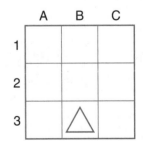

_____

b. In which row is the triangle located?

_____

SRB
144

125

LESSON 5·8 Reading and Writing Big Numbers

1. Each row in the place-value chart shows a number. Use words to write the name for each number below the chart.

	Billions				Millions				Thousands				Ones		
	100B	10B	1B	,	100M	10M	1M	,	100Th	10Th	1Th	,	100	10	1
a.			7	,	4	0	0	,	0	6	5	,	2	0	0
b.					5	1	,	8	0	0	,	0	0	0
c.		2	3	,	0	0	0	,	0	0	5	,	1	4	0
d.	1	2	3	,	4	5	6	,	7	8	9	,	0	1	2

a. *7 billion, 400 million, 65 thousand, 200* _____

b. _____

c. _____

d. _____

2. Use digits to write these numbers in the place-value chart below.

   a. 400 thousand, 500

   b. 208 million, 350 thousand, 600

   c. 16 billion, 210 million, 48 thousand, 715

   d. 1 billion, 1 million, 1 thousand, 1

	Billions				Millions				Thousands				Ones		
	100B	10B	1B	,	100M	10M	1M	,	100Th	10Th	1Th	,	100	10	1
a.															
b.															
c.															
d.															

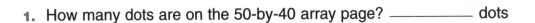

LESSON 5·8 How Much Are a Million and a Billion?

1. How many dots are on the 50-by-40 array page? _____ dots

2. How many dots would be on

   a. 5 pages? _____ dots

   b. 50 pages? _____ dots

   c. 500 pages? _____ dots

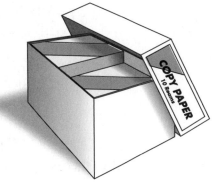

3. Each package of paper, or ream, contains 500 sheets. How many dots would be on the paper in

   a. 1 ream? (*Hint:* Look at Problem 2.)   _____ dots

   b. 10 reams? (1 carton)   _____ dots

   c. 100 reams? (10 cartons)   _____ dots

   d. 1,000 reams? (100 cartons)   _____ dots

4. Use digits to write these numbers in the place-value chart below.

   a. 999 thousand   b. 1,000 thousand   c. 999 million   d. 1,000 million

	Billions				Millions				Thousands				Ones		
	100B	10B	1B	,	100M	10M	1M	,	100Th	10Th	1Th	,	100	10	1
a.															
b.															
c.															
d.															

LESSON 5·8 Internet Users

The table below shows the estimated number of people who used the Internet in several different countries in 2004.

Internet Users in 2004

Country	Users
France	25,470,000
Greece	2,710,000
Hungary	2,940,000
Italy	25,530,000
Poland	10,400,000
Spain	13,440,000

Source: The World Factbook

1. Which of these countries had the most Internet users? _____

2. Which of these countries had the fewest Internet users? _____

3. About how many more Internet users were there in France than in Spain?

   Number model: _____

   Answer: _____

4. Write true or false.

   a. There were more Internet users in Greece, Poland, and Hungary combined than in Spain. _____

   b. There were about 6 times as many Internet users in France as there were in Hungary. _____

5. Why do you think the Internet usage data is rounded to the nearest 10,000 instead of an actual count?

   _____

   _____

6. The United States had about 62 times as many Internet users as Hungary. About how many Internet users were there in the United States?

   Number model: _____

   Answer: _____

LESSON 5·8

Math Boxes

---

1. a. Measure the line segment to the nearest $\frac{1}{4}$ inch.

R _____ ·S

About _____ inches

b. Draw a line segment that is half as long as the one above.

c. How long is the line segment you drew? About _____ inches

---

2. Estimate the product. Write a number model to show how you estimated.

a. 41 * 83

Number model:

_____

b. 75 * 32

Number model:

_____

---

3. Multiply. Use the partial-products method.

_____ = 46 * 98

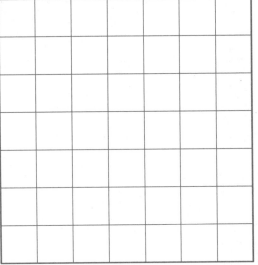

---

4. Write each number using digits.

a. seven hundred six thousandths

_____

b. three and four hundredths

_____

---

5. You have 75 apples. You need 8 apples to make a pie. How many pies can you bake?

○ **A.** 9

○ **B.** 8

○ **C.** 7

○ **D.** 6

LESSON 5·9   Place Value and Powers of 10

SRB
5

Millions	Hundred Thousands	Ten Thousands	Thousands	Hundreds	Tens	Ones
1,000,000			1,000	100		1
10 [100,000s]			10 [100s]	10 [10s]		10 [tenths]
		10 * 10 * 10 * 10		10 * 10		
	10^5		10^3	10^2		10^0

Fill in this place-value chart as follows:

1. Write standard numbers in Row 1.

2. In Row 2, write the value of each place to show that it is 10 times the value of the place to its right.

3. In Row 3, write the place values as products of 10s.

4. In Row 4, show the values as powers of 10. Use exponents. The exponent shows how many times 10 is used as a factor. It also shows how many zeros are in the standard number.

LESSON
5·9   **Math Boxes**

---

1. Estimate the sum. Write a number model to show how you estimated.

a. $1{,}254 + 8{,}902 + 2{,}877$

Number model:

_____

b. $12{,}645 + 7{,}302 + 15{,}297$

Number model:

_____

SRB
181

---

2. Which number sentence is true? Fill in the circle next to the best answer.

○ **A.** $5{,}800{,}000 = 58$ million

○ **B.** 62 million $> 3{,}100{,}000{,}000$

○ **C.** $10^3 = 1{,}000$

○ **D.** $100{,}000 = 10^2$

SRB
5   6

---

3. Draw lines to match each word to the correct pair or pairs of line segments.

perpendicular

parallel

intersecting

SRB
94   95

---

4. Complete.

Rule: _____

in	out
1.29	1.36
6.47	
	5.17
12.66	
7.93	8.00

SRB
162–166

---

5. Multiply. Use a paper-and-pencil algorithm.

$9 * 258 = $ _____

SRB
18   19

---

6. Which of the angles below has a measure of about 90 degrees? Circle it.

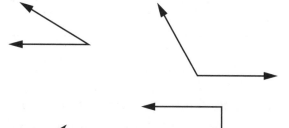

SRB
92   93

Date _____   Time _____

Evaluating Large Numbers

1. Round the attendance figures in the table to the nearest hundred-thousand.

2004 Major League Baseball Home Game Attendance for 10 Teams		
Team	Total Home Game Attendance	Attendance Rounded to the Nearest 100,000
Atlanta Braves	2,603,484	
Baltimore Orioles	2,682,917	
Boston Red Sox	2,650,063	
Cleveland Indians	2,616,940	
Colorado Rockies	2,737,838	
Los Angeles Dodgers	3,131,255	
New York Yankees	3,465,807	
Seattle Mariners	3,540,658	
St. Louis Cardinals	3,011,216	
Texas Rangers	2,352,397	

Source: Information Please—Baseball Attendance

2. How do you think attendance figures for major league baseball games are obtained?

_____

_____

_____

3. Do you think *exactly* 2,737,838 people were at the home games played by the Colorado Rockies? Explain your answer.

_____

_____

_____

4. You rounded the figures in the table above to the nearest hundred-thousand. Which teams have the same attendance figures based on these rounded numbers?

_____

_____

LESSON 5·10

Math Boxes

---

1. a. Measure the line segment to the nearest $\frac{1}{4}$ inch.

T ─────────────────────────────────── G

About _____ inches

b. Draw a line segment that is half as long as the one above.

c. How long is the line segment you drew?   About _____ inches

SRB
128

---

2. Estimate the product. Write a number model to show how you estimated.

a. 37 * 91

Number model:

_____

b. 53 * 17

Number model:

_____

SRB
184

---

3. Multiply. Use the partial-products method.

_____ = 43 * 89

SRB
18

---

4. Write fourteen and three-tenths using digits. Fill in the circle next to the best answer.

○ **A.** 14.03

○ **B.** 14.003

○ **C.** 14.310

○ **D.** 14.3

SRB
27 28

---

5. There are 60 trading cards. Each student gets 5 cards. How many students get trading cards?

_____ students

SRB
175

LESSON 5·11   **Traveling to Europe**

It is time to leave Africa. Your destination is Region 2—the continent of Europe. You and your classmates will fly from Cairo, Egypt to Budapest, Hungary. Before exploring Hungary, you will collect information about the countries in Region 2. You may even decide to visit another country in Europe after your stay in Budapest.

Use the World Tour section of your *Student Reference Book* to answer the questions.

1. Which country in Region 2 has

   a. the largest population?   _____

   country   population

   b. the smallest population?   _____

   country   population

   c. the largest area?   _____

   country   area

   d. the smallest area?   _____

   country   area

Use the Climate and Elevation of Capital Cities table on page 297.

2. From December to February, which capital in Region 2 has

   a. the warmest weather?   _____

   capital   country   temperatures

   b. the coolest weather?   _____

   capital   country   temperatures

   c. the greatest amount of rain?   _____

   capital   country   inches rainfall

   d. the least amount of rain?   _____

   capital   country   inches rainfall

Use the Population Data table on page 301.

3. Which country in Region 2 has

   a. the greatest percent of population ages 0–14?   _____

   country   percent

   b. the smallest percent of population ages 0–14?   _____

   country   percent

LESSON 5·11 Water, Water Everywhere

Solve each problem below. Record a number model for the problem using a letter for the unknown. You may want to use two number models for some of the problems. Then write a summary number model with your answer in place of the letter.

SRB
178A
178B

1. The world's largest lake is the Caspian Sea, with an area of about 143,200 square miles. The second largest lake, Lake Superior, has an area of about 31,320 square miles. What is the approximate total area of both lakes?

   Answer: About _____ square miles

   _____
   (number model with unknown)

   _____
   (number model with answer)

2. The Nile River in Egypt is about 4,132 miles long. The longest river in the United States, the Missouri River, stretches about 2,540 miles. How much longer is the Nile River than the Missouri River?

   Answer: About _____ miles

   _____
   (number model with unknown)

   _____
   (number model with answer)

3. To grow a single orange, it takes about 13.8 gallons of water. A tomato is made of 95% water, but takes only 3 gallons of water to grow it. How much more water is needed to grow an orange than a tomato?

   Answer: About _____ gallons

   _____
   (number model with unknown)

   _____
   (number model with answer)

134A

LESSON 5·11   **Water, Water Everywhere** *continued*

4. The average depth of the ocean is 4,267 meters. The deepest spot, Challenger Deep in the Mariana Trench near Guam, is about 11,030 meters below the surface. How much deeper is Challenger Deep than the average depth of the ocean?

Answer: About _____ meters

_____
(number model with unknown)

_____
(number model with answer)

5. The total annual rainfall for the three wettest inhabited places in the world is 1,416 inches. In the wettest place, Cherrapunji, India, it rains about 498 inches per year. In the second wettest place, Mawsynram, India, it rains about 467 inches per year. About how many inches per year does it rain in the third wettest place, Waialeale, Hawaii?

Answer: About _____ in.

_____
(number model(s) with unknown)

_____
(number model(s) with answer)

6. Alaska, the biggest state in the United States, has more miles of rivers and streams than any other state. The next four highest ranked states are California with 211,513 miles; Texas with 191,228 miles; Montana with 176,750 miles; and Nevada with 143,750 miles. The top five states have 1,088,241 miles of rivers and streams. How many miles of rivers and streams does Alaska have?

Answer: About _____ miles

_____
(number model(s) with unknown)

_____
(number model(s) with answer)

Math Boxes

---

1. Estimate the sum. Write a number model to show how you estimated.

a. 799 + 11,304 + 48,609

Number model:

_____

b. 4,382 + 6,911 + 7,035

Number model:

_____

181

---

2. Write <, >, or = to make each number sentence true.

a. 356,789 _____ 354,999

b. 670,000 _____ 67,000,000

c. 62 million _____ 9,700,000

d. 105,000,000 _____ 15,500,000

e. 10^4 _____ 1,000

5   6

---

3. **a.** Draw a pair of parallel line segments.

b. Draw a pair of perpendicular line segments.

94   95

---

4. Complete.

Rule: _____

in	out
6.46	6.58
3.08	
	11.34
25.25	
	100.1
63.09	63.21

162–166

---

5. Multiply. Use a paper-and-pencil algorithm.

7 * 208 = _____

18   19

---

6. Which of the angles below has a measure less than 90 degrees? Circle it.

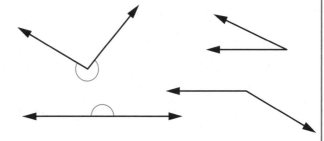

93

LESSON 5·12   **Math Boxes**

---

1. There are 240 chairs to set up for the concert. Each row has 40 chairs in it. How many rows are there?

_____ rows

SRB
21

2. The senior class at Rees High School raised $1,895 for five charities in the community. The money will be shared equally. How much money will each charity receive?

_____

SRB
22 23

---

3. Look at the grid below.

a. In which column is the star located?

_____

b. In which row is the star located?

_____

SRB
144

4. Draw a right angle with vertex *K*.

SRB
93

---

5. Divide.

a. 72 / 9 = _____

b. 720 / 90 = _____

c. 7,200 / 900 = _____

d. _____ = 42 / 7

e. _____ = 420 / 7

f. _____ = 4,200 / 7

g. 28 / 4 = _____

h. 28,000 / 40 = _____

i. 28,000 / 400 = _____

j. _____ = 24 / 6

k. _____ = 2,400 / 60

l. _____ = 24,000 / 6

SRB
20 21

LESSON 6·1 | **Math Boxes**

---

1. Measure each line segment to the nearest millimeter.

a.

R   S

About _____ cm _____ mm

b.

C   S

About _____ cm _____ mm

SRB 128

---

2. Round 409,381,886 to the nearest

a. hundred.

_____

b. ten thousand.

_____

c. ten million.

_____

d. hundred million.

_____

SRB 182 183

3. Multiply. Use a paper-and-pencil method.

_____ = 86 * 29

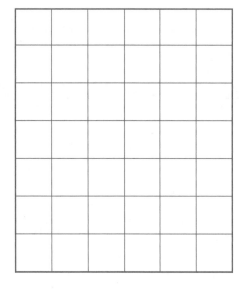

SRB 18 19

---

4. Complete.

a. $10^2 =$ _____

b. $10^{\square} = 10 * 10 * 10 * 10$

c. $1{,}000 = 10^{\square}$

d. 10 to the ninth power =

_____

SRB 5

5. Circle $\frac{5}{6}$ of the squares.

SRB 44

LESSON 6·1 Multiplication/Division Number Stories

SRB
178A
178B

Fill in each Multiplication/Division Diagram. Then write a number model with a variable for the unknown. Be sure to include a unit with your answer. Solve the problem and write a summary number model.

1. The profit from the Maple Street lemonade stand was $36. Four girls will share this amount equally. What will each girl's share be?

girls	dollars per girl	total dollars

Number model with unknown: _____

Answer: _____ Summary number model: _____

2. Sheila has 57 pictures to put in her photo album. She can put 6 pictures on each page. How many pages will be completely filled up when she is finished?

pages	pictures per page	total pictures

Number model with unknown: _____

Answer: _____ Summary number model: _____

3. Reuben walks a total of 35 blocks going to and from school each week. He always walks the same route. How many blocks does he walk each day?

days	blocks per day	total blocks

Number model with unknown: _____

Answer: _____ Summary number model: _____

LESSON 6·1 — Multiplication/Division Number Stories *cont.*

4. Hassan is helping his teacher put 8 centimeter cubes into each paper cup for a math project. How many paper cups can he fill if there are 79 cubes?

paper cups	cubes per paper cup	total cubes

Number model with unknown: _____

Answer: _____ Summary number model: _____

Try This

5. Mr. Henning's fourth-grade class is planning a field trip to see a play. The bus will cost $100, and the tickets will cost $125. The 25 students will share the total cost equally. How much will each student pay for the field trip?

_____	_____ per _____	total _____

Number model with unknown: _____

Answer: _____ Summary number model: _____

6. Last year, Martina sold 73 boxes of cookies for her club. This year, she sold three times as many. If she collected $876 this year, how much did each box cost?

_____	_____ per _____	total _____

Number model(s) with unknown: _____

Answer: _____

Summary number model(s): _____

LESSON 6·1 Extended Multiplication Facts

SRB
17

1. 9 * 5 = _____

9 * 50 = _____

90 * 5 = _____

90 * 50 = _____

900 * 5 = _____

90 * 500 = _____

2. 8 * 7 = _____

8 * 70 = _____

80 * 7 = _____

80 * 70 = _____

800 * 7 = _____

80 * 700 = _____

3. 4 * 9 = _____

4 * 90 = _____

40 * 9 = _____

40 * 90 = _____

400 * 9 = _____

40 * 900 = _____

4. 6 * _____ = 18

60 * _____ = 180

60 * _____ = 1,800

_____ * 60 = 180

_____ * 600 = 1,800

30 * _____ = 18,000

5. _____ * 8 = 48

_____ * 80 = 480

_____ * 80 = 4,800

60 * _____ = 480

6 * _____ = 4,800

6 * _____ = 48,000

6. 8 * _____ = 24

8 * _____ = 2,400

80 * _____ = 2,400

_____ * 30 = 240

_____ * 3 = 240

_____ * 300 = 240,000

LESSON 6·2 — Math Boxes

1. There are 32 students in the class. A yearbook page can show 8 student photos. How many pages are needed to include all the student photos?

pages	photos per page	total photos

Number model with unknown:

_____

Answer: _____ pages

Summary number model:

_____

2. Solve each open sentence.

a. $24 = a * (5 + 1)$ $a =$ _____

b. $54 / 6 = 81 / b$ $b =$ _____

c. $(c + 4) / 3 = 7$ $c =$ _____

d. $m - 3.87 = 7.49$ $m =$ _____

e. $0.98 + 4.83 = f + 4.35$ $f =$ _____

3. Use a paper-and-pencil algorithm to add or subtract.

a.
$$\begin{array}{r} 0.85 \\ + 0.53 \\ \hline \end{array}$$

b.
$$\begin{array}{r} 0.64 \\ + 1.73 \\ \hline \end{array}$$

c.
$$\begin{array}{r} 12.38 \\ - 1.09 \\ \hline \end{array}$$

d.
$$\begin{array}{r} 3.05 \\ - 0.67 \\ \hline \end{array}$$

4. Complete.

a. 670 cm = _____ m

b. 4,800 cm = _____ m

c. 916 cm = _____ m _____ cm

d. 18 m = _____ cm

5. Name a fraction equivalent to $\frac{1}{2}$. Circle the best answer.

A. $\frac{3}{4}$ B. $\frac{8}{9}$

C. $\frac{5}{10}$ D. $\frac{3}{5}$

141

LESSON 6·2   Solving Division Problems

For Problems 1–6, fill in the multiples-of-10 list if it is helpful. If you prefer to solve the division problems in another way, show your work.

1. José's class baked 64 cookies for the school bake sale. Students put 4 cookies in each bag. How many bags of 4 cookies did they make?

   10 [4s] = _____   Number model with unknown: _____

   20 [4s] = _____   Answer: _____ bags

   30 [4s] = _____   Summary number model: _____

   40 [4s] = _____

   50 [4s] = _____

2. The community center bought 276 cans of soda for a picnic. How many 6-packs is that?

   10 [6s] = _____   Number model with unknown: _____

   20 [6s] = _____   Answer: _____ 6-packs

   30 [6s] = _____   Summary number model: _____

   40 [6s] = _____

   50 [6s] = _____

3. Each lunch table at Johnson Elementary School seats 5 people. How many tables are needed to seat 191 people?

   10 [5s] = _____   Number model with unknown: _____

   20 [5s] = _____   Answer: _____ tables

   30 [5s] = _____   Summary number model: _____

   40 [5s] = _____

   50 [5s] = _____

LESSON 6·2

Solving Division Problems *continued*

4. The preschool held a tricycle parade. Trent counted 135 wheels.
   How many tricycles is that?

   10 [3s] = _____   Number model with unknown: _____

   20 [3s] = _____   Answer: _____ tricycles

   30 [3s] = _____   Summary number model: _____

   40 [3s] = _____

   50 [3s] = _____

---

5. How many 8s are there in 248?

   10 [8s] = _____   Number model with unknown: _____

   20 [8s] = _____   Answer: _____

   30 [8s] = _____   Summary number model: _____

   40 [8s] = _____

   50 [8s] = _____

---

6. How many 7s are in 265?

   10 [7s] = _____   Number model with unknown: _____

   20 [7s] = _____   Answer: _____

   30 [7s] = _____   Summary number model: _____

   40 [7s] = _____

   50 [7s] = _____

Partial-Quotients Division Algorithm

1. There are 6 pencils in each pack. How many packs can be made from 96 pencils?

Number model with unknown:

_____

Answer: _____ packs

How many pencils are left over? ____ pencils

Summary number model:

_____

2. Phil has $79 to purchase books. Books cost $7 each. How many books can Phil buy?

Number model with unknown:

_____

Answer: _____ books

How many dollars are left over? ____ dollars

Summary number model:

_____

3. There are 184 plants to be put into pots. Each pot can hold 8 plants. How many pots are needed?

Number model with unknown:

_____

Answer: _____ pots

How many plants are left over? ____ plants

Summary number model:

_____

4. A waiter distributed 1,325 drinking straws evenly among 9 dispensers. How many straws went in each dispenser?

Number model with unknown:

_____

Answer: _____ straws

How many straws were left over? ____ straws

Summary number model:

_____

LESSON 6·3

Partial-Quotients Division Algorithm *cont.*

Divide.

5. 3)87‾

Answer: _____

6. 1,081 ÷ 7

Answer: _____

Try This

7. A factory has 372 boxes of shirts to distribute evenly among 12 stores. Each box holds 15 shirts. How many shirts will each store receive?

Number model(s) with unknown:

_____

_____

Answer: _____ shirts

How may shirts are left over? _____ shirts

Summary number model(s):

_____

_____

8. There are _____ players in the league. (Write a number greater than 100.)

There are _____ players on each team. (Write a number between 3 and 9.)

How many teams can be made?

Number model with unknown:

_____

Answer: _____ teams

How many players are left over? _____ players

Summary number model:

_____

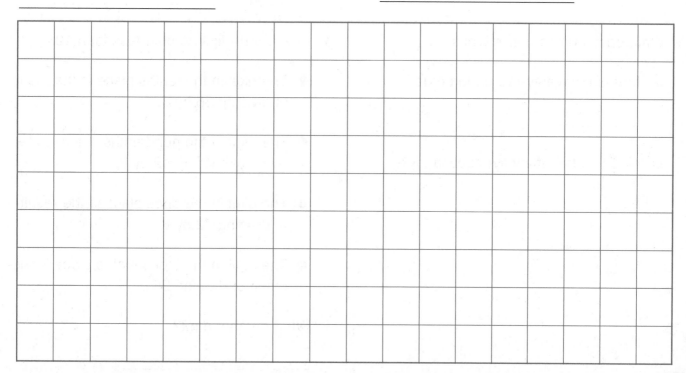

LESSON 6·3 Place Value in Decimals

SRB
30 31

1. Write these numbers in order from smallest to largest.

1.26   0.58   1.09   1.091   0.35

_____ smallest

_____

_____

_____

_____ largest

2. A number has

6 in the tenths place,
4 in the ones place,
5 in the hundredths place, and
9 in the tens place.

Write the number.

____ ____ . ____ ____

3. Write the smallest number you can make with the following digits:

3   6   4   7   2

_____

4. What is the value of the digit 4 in the numerals below?

a. 37.48 _____

b. 49.08 _____

c. 0.942 _____

d. 1.664 _____

5. Write each number using digits.

a. four and seventy-two hundredths

_____

b. nine hundred thirty-five thousandths

_____

6. I am a four-digit number less than 10.

◆ The digit in the tenths place is the result of dividing 36 by 4.

◆ The digit in the hundredths place is the result of dividing 42 by 7.

◆ The digit in the ones place is the result of dividing 72 by 8.

◆ The digit in the thousandths place is the result of dividing 35 by 5.

What number am I?

____ . ____ ____ ____

LESSON 6·3   **Math Boxes**

1. Measure each line segment to the nearest millimeter.

   a. _____
   P   S

   About _____ cm _____ mm

   b. _____
   A   B

   About _____ cm _____ mm

   SRB 128

2. Round 5,906,245 to the nearest

   a. million.

   _____

   b. ten thousand.

   _____

   c. thousand.

   _____

   d. hundred.

   _____

   SRB 182 183

3. Multiply. Use a paper-and-pencil method.

   _____ = 58 * 52

   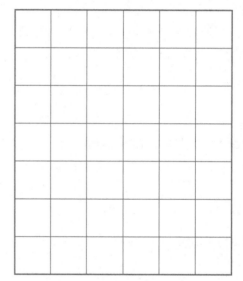

   SRB 18 19

4. Complete.

   a. $10^4 =$ _____

   b. $10^{\boxed{}} = 10 * 10 * 10 * 10 * 10$

   c. $100 = 10^{\boxed{}}$

   d. 10 to the seventh power =

   _____

   SRB 5

5. Circle $\frac{1}{2}$ of the squares.

   SRB 44

147

LESSON 6·4 Interpreting Remainders

For each number story:

◆ Draw a picture.
◆ Write a number model with a letter for the unknown.
◆ Use a division algorithm to solve the problem.
◆ Decide what to do about the remainder.
◆ Record the answer and write a summary number model.

SRB
179

1. Jackson is buying balloons for a party. Balloons cost $6 per bunch. How many bunches can he buy with $75?

   Picture:

   Number model with unknown:

   _____

   Answer: _____ bunches

   Summary number model:

   _____

   What did you do about the remainder? Circle the answer.

   A. Ignored it

   B. Reported it as a fraction or decimal

   C. Rounded the answer up

   Why? _____

   _____

2. Rosa is buying boxes to hold all 128 of her CDs. Each box holds 5 CDs. How many boxes are needed to store all of her CDs?

   Picture:

   Number model with unknown:

   _____

   Answer: _____ boxes

   Summary number model:

   _____

   What did you do about the remainder? Circle the answer.

   A. Ignored it

   B. Reported it as a fraction or decimal

   C. Rounded the answer up

   Why? _____

   _____

LESSON 6·4 Interpreting Remainders *continued*

3. Lateefah won 188 candy bars in a raffle. She decided to share them equally with 7 of her classmates and herself. How many candy bars did each person receive?

Picture:

Number model with unknown:

_____

Answer: _____ candy bars

Summary number model:

_____

What did you do about the remainder? Circle the answer.

A. Ignored it

B. Reported it as a fraction or decimal

C. Rounded the answer up

Why? _____

_____

_____

Try This

Write each answer as a mixed number by rewriting the remainder as a fraction.

4. $2\overline{)27}$ $\overset{13\ R1}{}$ _____

5. $10\overline{)883}$ $\overset{88\ R3}{}$ _____

6. $16\overline{)252}$ $\overset{15\ R12}{}$ _____

Write each answer as a decimal.

7. $39 \div 2 = 19\ R1$ _____

8. $183 \div 12 = 15\ R3$ _____

9. $2{,}067 \div 5 = 413\ R2$ _____

LESSON 6·4 **Math Boxes**

---

1. Joe ordered 72 plants for his patio garden. Each pot holds 4 plants. How many pots are needed to hold all of the plants?

pots	plants per pot	total plants

Number model with unknown:

_____

Answer: _____

Summary number model:

_____

178A
178B

2. Solve each open sentence.

a. $(6 + 9) + (3 * A) = 30$ $A =$ _____

b. $24 \div 8 = 21 \div B$ $B =$ _____

c. $72 = (2 * C) * 9$ $C =$ _____

d. $6.2 + 0.79 = D$ $D =$ _____

e. $8.91 - E = 2.72$ $E =$ _____

35–37
148

---

3. Use a paper-and-pencil algorithm to add or subtract.

a.
```
   0.37
 + 0.26
```

b.
```
   2.9
 + 5.01
```

c.
```
   6.79
 - 6.55
```

d.
```
   7.80
 - 3.65
```

34–37

---

4. How many centimeters are in 12 meters? Circle the best answer.

A. 0.12   B. 1.2

C. 120   D. 1,200

129

5. Circle the fractions equivalent to $\frac{1}{2}$.

$\frac{8}{16}$ $\frac{5}{6}$ $\frac{6}{12}$

$\frac{2}{3}$ $\frac{12}{24}$ $\frac{8}{15}$

51

LESSON 6·5

Math Boxes

---

1. Insert parentheses to make each number sentence true.

   a. 15 + 5 * 6 = 120

   b. 7 + 9 * 2 = 25

   c. 77 = 1 + 6 * 6 + 5

   SRB
   150

---

2. Draw a line segment that is 2 inches long. Mark and label the following inch measurements on the line segment:

   $\frac{1}{2}$, $\frac{3}{4}$, 1, 1$\frac{1}{2}$, and 2

   SRB
   128

---

3. The Sports Boosters raised $908 at their annual chili supper. Four athletic teams will share the money equally.

   How much money will each team receive?

   Number model with unknown:

   _____

   Answer: _____

   Summary number model:

   _____

   SRB
   22   23

---

4. Multiply with a paper-and-pencil algorithm.

   66 * 62 = _____

   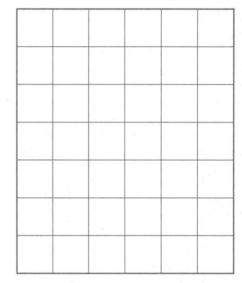

   SRB
   18   19

---

5. Complete.

   a. 9 m = _____ cm

   b. 1,500 cm = _____ m

   c. 350 cm = _____ m

   d. 458 cm = _____ m _____ cm

   e. 3.2 m = _____ cm

   SRB
   129

---

6. **a.** Shade $\frac{1}{2}$ of the square.

   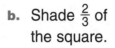

   b. Shade $\frac{2}{3}$ of the square.

   SRB
   44

LESSON 6·5   **Making a Full-Circle Protractor**

There are 360 marks around the circle. They divide the edge of the circle into 360 small spaces. Twelve of the marks are longer than the rest. They are in the same positions as the 12 numbers around a clock face. Your teacher will tell you how to label the 12 large marks on the circle.

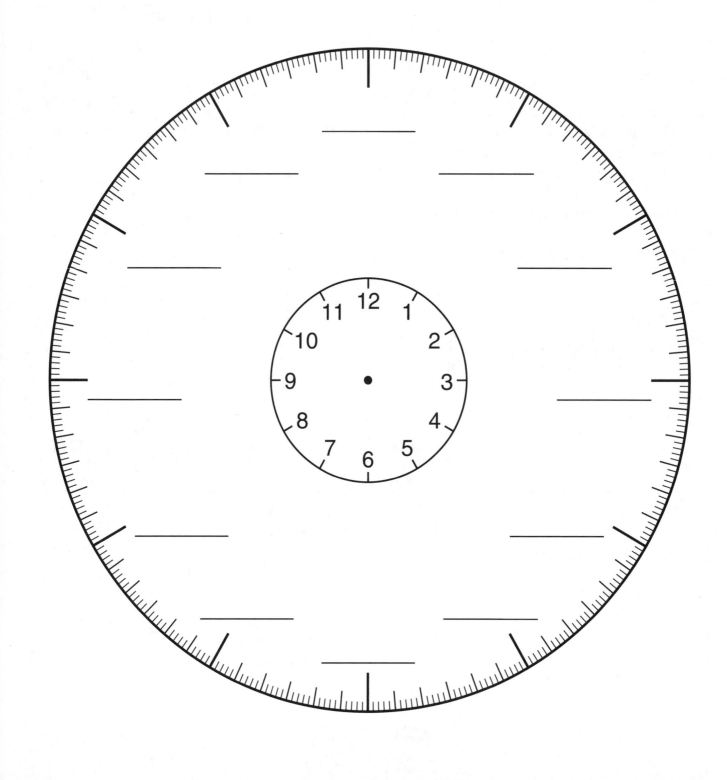

LESSON 6·5   Clock Angles

Use the clock below and the full-circle protractor on journal page 152 to help you answer the questions.

1. How many minutes and how many degrees does the *minute hand* move

   a. from 3:00 to 4:00?   _____ minutes   _____ °

   b. from 7:00 to 7:45?   _____ minutes   _____ °

   c. from 8:15 to 8:45?   _____ minutes   _____ °

   d. from 6:30 to 6:50?   _____ minutes   _____ °

   e. from 5:15 to 5:30?   _____ minutes   _____ °

   f. from 1:00 to 1:10?   _____ minutes   _____ °

   g. from 12:00 to 12:05?   _____ minutes   _____ °

   h. from 5:00 to 5:01?   _____ minutes   _____ °

Try This

2. How many degrees does the *hour hand* move

   a. in 1 hour? _____ °

   b. in $\frac{1}{2}$ hour? _____ °

   c. in 10 minutes? _____ °

3. Explain how you solved Problem 2c.

   _____

   _____

   _____

   _____

Population Bar Graph

The table below shows the percent of the population (number of people out of 100) who are 14 years old or younger in the Region 2 countries.

Country	Percent of Population Ages 0–14
France	19
Greece	15
Hungary	16
Iceland	23
Italy	14
Netherlands	18
Norway	20
Poland	18
Spain	15
United Kingdom	19

1. Make a bar graph to display the information given in the table above.

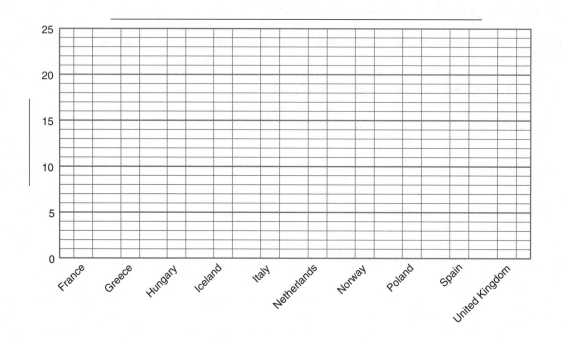

2. Why might it be important to know what percent of the population of a country is 0 through 14 years of age?

_____

_____

Elapsed Time

Record the times on each pair of clocks. Then record the elapsed time.

1.　　Start　　　　　End

_____　_____

_____

2.　　Start　　　　　End

_____　_____

_____

3.　　Start　　　　　End

_____　_____

_____

4.　　Start　　　　　End

_____　_____

_____

Record how much time has passed between the start time and the end time.

5. Start 11:00 A.M.　　　　Elapsed time: _____
　　 End 4:30 P.M.

6. Start 2:20 P.M.　　　　 Elapsed time: _____
　　 End 6:35 P.M.

7. Start 9:12 A.M.　　　　 Elapsed time: _____
　　 End 11:43 P.M.

LESSON 6·5   **Elapsed Time** *continued*

Read the time on each clock. What time will it be in 50 minutes?

8.

9.

10.

_____   _____   _____

For each time, record what time it will be in 1 hour and 20 minutes.

11. 11:00 A.M.   **12.** 6:45 P.M.   **13.** 9:53 P.M.

_____   _____   _____

Read the time on each clock. What time was it 30 minutes ago?

14.

15.

16.

_____   _____   _____

For each time, record what time it was 2 hours and 15 minutes ago.

11. 10:15 A.M.   **12.** 2:05 P.M.   **13.** 1:12 A.M.

_____   _____   _____

LESSON 6·6 Measuring Angles

Use your full-circle protractor to measure each angle.

SRB
92

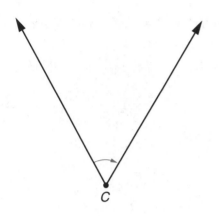

1. ∠C measures _____ °.

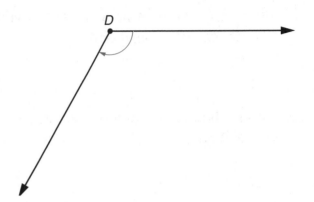

2. ∠D measures _____ °.

Try This

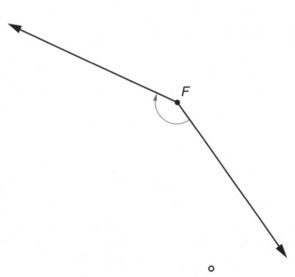

3. ∠F measures _____ °.

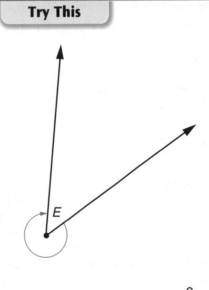

4. ∠E measures _____ °.

5. Without using your full-circle protractor, give the measure of the reflex angle in Problem 3 (the part not marked by the blue arrow). Explain your answer.

_____

_____

_____

LESSON 6·6

Math Boxes

1. Ms. Kawasaki's fourth grade class made a circle graph to show students' favorite days of the week.

   a. Which day of the week is the least favorite in Ms. Kawasaki's classroom?

   _____

   b. About what fraction of the students prefer Saturday?

   _____

Favorite Day of the Week

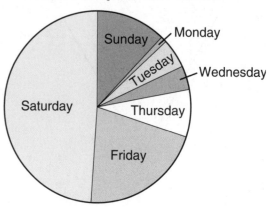

2. Juan talked on the phone an average of 34 minutes per week for 1 whole year. About how many minutes did Juan spend on the phone in 1 year?

   Number model with unknown:

   _____

   Answer: _____ minutes

   Summary number model:

   _____

3. Divide with a paper-and-pencil algorithm. Write the remainder as a fraction.

   883 / 7 = _____

4. Write <, >, or = to make each number sentence true.

   a. 420,000,000 _____ four hundred twenty million

   b. 65,000,000 _____ 92,000,000

   c. four hundred thousand _____ 10^4

   d. 10^2 _____ 1,000

5. For this spinner, what color would you be *most likely* to land on?

   _____

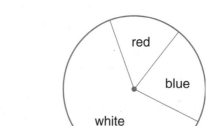

156

LESSON 6·7 Drawing and Measuring Angles

Math Message

Use a straightedge to draw the following angles. Do **not** use a protractor.

∠A: any angle that
measures less
than 90°

∠B: any angle that measures
more than 90° and less
than 180°

∠C: any angle that
measures more
than 180°

∠A is called an **acute angle.**   ∠B is called an **obtuse angle.**   ∠C is called a **reflex angle.**

Measuring Angles with a Protractor

Write whether the angle is *acute* or *obtuse*. Then measure it as accurately as you can.

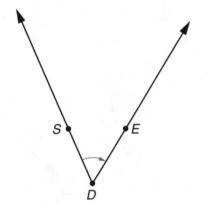

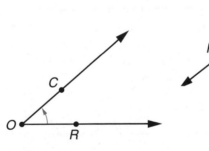

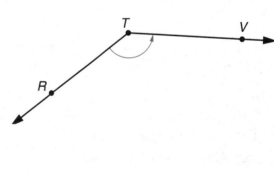

∠SDE is _____.

∠COR is _____.

∠RTV is _____.

∠SDE measures _____°.

∠COR measures _____°.

∠RTV measures _____°.

LESSON 6·7 **Drawing Angles**

1. Draw a 35° angle, using line segment *GH* as one of its sides.

2. Draw a 150° angle, using ray *CD* as one of its sides.

3. Draw a 60° angle, using ray *EF* as one of its sides.

4. Draw a 15° angle, using ray *AB* as one of its sides.

Try This

5. Draw a 330° angle, using ray *IJ* as one of its sides.

LESSON 6·7

Math Boxes

---

1. Insert parentheses to make each number sentence true.

   a. $12 = 15 - 2 + 1$

   b. $66 - 16 * 4 = 200$

   c. $49 = 4 + 3 * 42 / 6$

2. Draw a line segment that is 2 inches long. Mark and label the following inch measurements on the line segment:

   $\frac{1}{4}, \frac{3}{4}, 1, 1\frac{1}{4}$ and $1\frac{1}{2}$

3. Six classrooms collected newspapers for one week. If they collected a total of 582 newspapers by the end of the week, on average about how many newspapers did each class collect?

   Number model with unknown:

   _____

   Answer: _____ newspapers

   Summary number model:

   _____

4. Multiply with a paper-and-pencil algorithm.

   $67 * 34 =$ _____

   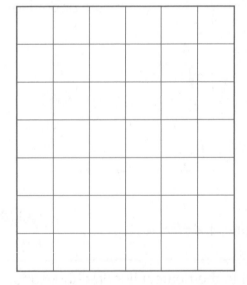

5. How many centimeters are in 9.7 meters? Circle the best answer.

   A. 907

   B. 900.7

   C. 970

   D. 9,700

6. Circle the square that has $\frac{1}{3}$ shaded.

   A.   **B.**

LESSON 6·8 — Math Boxes

1. Name the ordered number pair for each point plotted on the coordinate grid.

A (_____, _____)

B (_____, _____)

C (_____, _____)

D (_____, _____)

E (_____, _____)

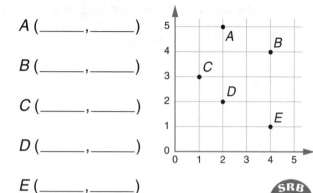

SRB
144

2. Complete the "What's My Rule?" table and state the rule.

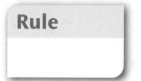

Rule

in	out
3.6	2.1
10	8.5
7.2	
	4.9

SRB
162–166

3. ∠EDF is _____ (acute or obtuse).

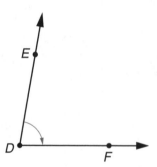

Measure of ∠EDF = _____°.

SRB
93
142 143

4. Cross off the names that do not belong in the name-collection box below.

32
81 − 49
9 * 4
(5 * 6) + 2
98 ÷ 3
10 + 15 + 7

SRB
149

5. Round to the nearest hundred-thousand.

a. 9,540,234   _____

b. 37,609,034   _____

c. 78,291,554   _____

d. 290,696,332 _____

SRB
182 183

6. Fill in the missing fractions on the number line.

0   1

_____ _____ _____

SRB
316

LESSON 6·8 A Map of the Island of Ireland

Bantry	B-1	Dublin	F-4	Lahinch	B-4	Omagh	E-7
Belfast	F-7	Dundalk	F-6	Larne	F-7	Tralee	B-2
Carlow	E-3	Galway	C-4	Limerick	C-3	Tuam	C-5
Castlebar	B-6	Gort	C-4	Mullingar	E-5	Westport	B-5
Derry	E-8	Kilkee	B-3	Navan	E-5	Wicklow	F-4

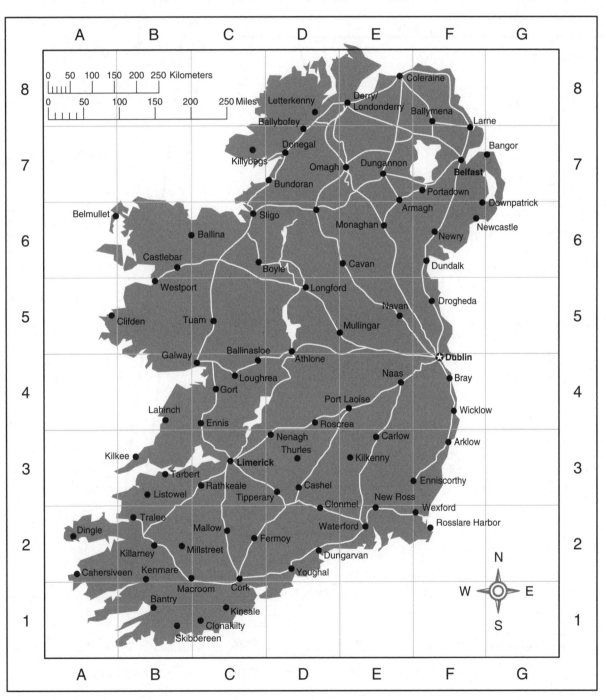

LESSON 6·8 A Campground Map

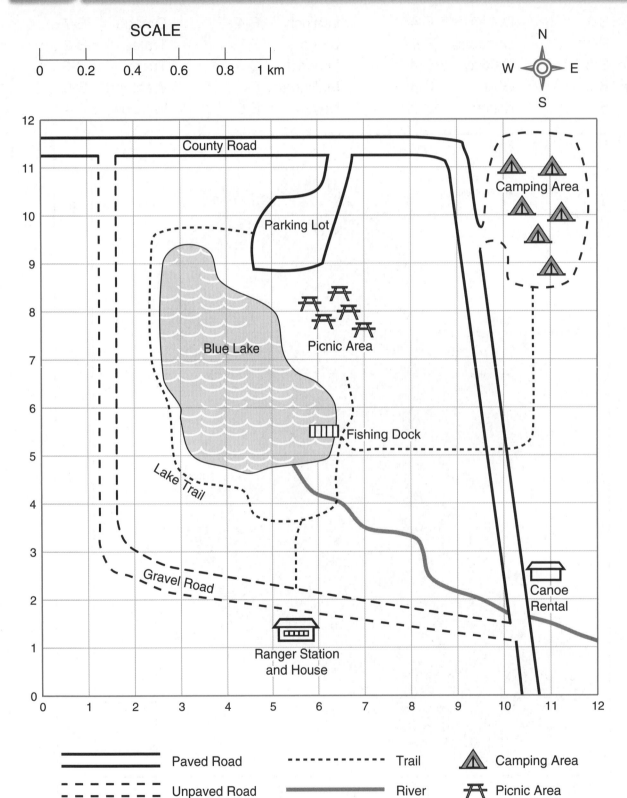

SCALE

0   0.2   0.4   0.6   0.8   1 km

N
W — ◇ — E
S

County Road

Parking Lot

Blue Lake

Picnic Area

Fishing Dock

Lake Trail

Gravel Road

Ranger Station and House

Camping Area

Canoe Rental

———— Paved Road   ·········· Trail   ▲ Camping Area

— — — Unpaved Road   ——— River   ⊼ Picnic Area

LESSON 6·8 Finding Distances on a Map

Use the campground map on journal page 162 to complete the following:

1. Suppose you hiked along the lake trail from the fishing dock to the parking lot. Estimate the distance you hiked.

   About _____ km

2. The ranger made her hourly check. She started at the ranger station. She drove northwest and then north on Gravel Road to County Road. She turned east onto County Road and drove past the parking lot and the camping area. After she passed the canoe rental, she turned right onto Gravel Road and drove back to the ranger station. About what distance did she drive?

   About _____ km

3. Estimate the distance around Blue Lake.

   About _____ km

4. You are planning to hike from the camping area to the parking lot. You will stay on the roads or trails. You want to hike at least 5 kilometers.

   a. Plan your route. Then draw it on the map with a colored pencil or crayon.

   b. Estimate the distance.

   About _____ km

5. Use the ordered number pairs to locate each item on the map. Mark a dot at the location. Next to the dot, write the letter given for the feature.

Campground Features Chart		
	Location	Letter
parked car	(5,9)	C
boat	$(3\frac{1}{2},8)$	B
swing set	(8,11)	S
hikers	(10.5,6.5)	H
farmhouse	$(\frac{1}{2},7)$	F

LESSON 6·8   **Dartboard Angles**

A regulation dartboard is made up of 20 equal sectors. The measure of each sector's angle is 18 degrees.

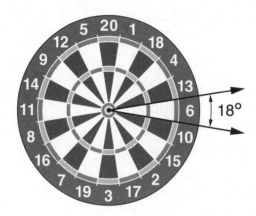

Solve Problems 1–3 without using your protractor. After you have solved each problem, write a number model with a letter for the unknown to show how you found your answer.

1. Draw ∠*ABC* around sectors 14, 9, 12, 5, 20, and 1. What is the measure of ∠*ABC*?

Measure of ∠*ABC* = _____°

Number model with unknown: _____

Dartboard Angles *continued*

2. The measure of angle *DFE* = 54° and the measure of angle *DFG* = 252°. What is the measure of reflex angle *EFG*?

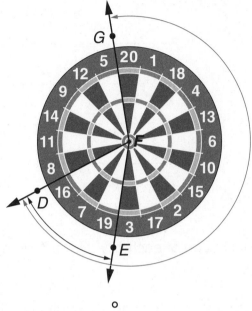

Measure of reflex angle *EFG* = _____

Number model with unknown: _____

3. Angle *RST* is a right angle. Use your straightedge to draw ray *SM* so that the measure of ∠*MST* = 54° and ∠*RSM* is an acute angle. Then find the measure of ∠*RSM*.

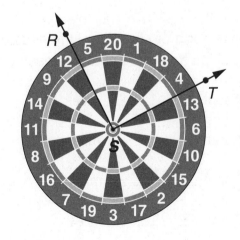

Measure of ∠*RSM* = _____

Number model with unknown: _____

LESSON 6·9 Locating Places on Regional Maps

Use the maps on pages 282–293 in the *Student Reference Book* to answer Problems 1–3.

1. Record the continent in which each city is located.

   a. Pretoria, South Africa (Region 1)   _____

   b. London, England (Region 2)   _____

   c. La Paz, Bolivia (Region 3)   _____

   d. Dhaka, Bangladesh (Region 4)   _____

   e. Washington, D.C., USA (Region 5)   _____

2. Find the approximate latitude and longitude of each city. Record the degrees and circle the correct direction.

   a. Pretoria, South Africa   latitude _____ °N or °S; longitude _____ °E or °W

   b. London, England   latitude _____ °N or °S; longitude _____ °E or °W

   c. La Paz, Bolivia   latitude _____ °N or °S; longitude _____ °E or °W

   d. Dhaka, Bangladesh   latitude _____ °N or °S; longitude _____ °E or °W

   e. Washington, D.C., USA   latitude _____ °N or °S; longitude _____ °E or °W

3. Each degree of latitude that you travel north or south from the equator is equal to about 70 miles. About how many miles from the equator is each city?

   a. Pretoria, South Africa   About _____ miles

   b. London, England   About _____ miles

   c. La Paz, Bolivia   About _____ miles

   d. Dhaka, Bangladesh   About _____ miles

   e. Washington, D.C., USA   About _____ miles

LESSON 6·9 **Math Boxes**

---

1. Cindy received $40 from her aunt and uncle. She drew a circle graph to show how she will use the money.

   a. How much will she save?

   _____

   b. How much will be spent on clothes?

   _____

   c. On movies?

   _____

   Cindy's Money

   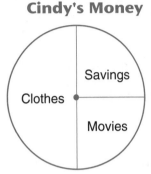

---

2. Mrs. Moy's students are folding paper cranes for an art project. Each of her 27 students is assigned to make at least 15 paper cranes. What is the least number of cranes the class will have for the project?

   Number model with unknown:

   _____

   Answer: _____ paper cranes

   Summary number model:

   _____

   18 19

3. Divide with a paper-and-pencil algorithm. Write the remainder as a fraction.

   598 / 3 = _____

   22 23
   179

---

4. Which number sentence is true? Circle the best answer.

   A. $33,000,000 < 33,000$

   B. $5,200,000 > 9$ million

   C. $10^4 = 10,000$

   D. six hundred thousand $= 10^6$

   5 6

5. For this spinner, which color would you be *least likely* to land on?

   _____

   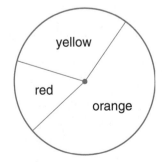

   80 84

LESSON 6·10

Partial-Quotients Division

1. Raul baked 96 cupcakes. He wants to divide them evenly among 3 bake sale tables. How many cupcakes should he put on each table?

   Number model with unknown:

   _____

   Answer: _____ cupcakes

   How many cupcakes will be left over?

   _____ cupcakes

   Summary number model:

   _____

2. The library has boxes to store 132 videotapes. Each box holds 8 tapes. How many boxes will be completely filled?

   Number model with unknown:

   _____

   Answer: _____ boxes

   How many videotapes will be left over?

   _____ videotapes

   Summary number model:

   _____

LESSON
6·10

Partial-Quotients Division *continued*

3. The teacher divided 196 note cards evenly among 14 students. How many note cards did each student get?

Number model with unknown:

_____

Answer: _____ note cards

How many note cards were left over?

_____ note cards

Summary number model:

_____

4. 18)864   Answer: _____

5. 509 ÷ 37 = _____

Try This

6. 4,872 / 24 = _____

7. 3,315 ÷ 36 = _____

8. The principal divided 462 boxes of markers evenly among 14 classrooms. There are 12 markers per box. How many markers does each classroom get?

Number model(s) with unknown:

_____

Answer: _____ markers

Summary number model(s):

_____

Math Boxes

---

1. Name the ordered number pair for each point plotted on the coordinate grid.

A (_____ , _____)

B (_____ , _____)

C (_____ , _____)

D (_____ , _____)

E (_____ , _____)

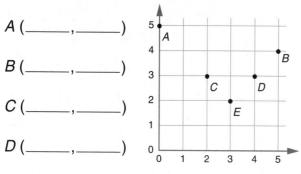

SRB
144

---

2. Complete the "What's My Rule?" table and state the rule.

in	out
3.66	7.04
0.42	3.80
8.73	
	12.66

SRB
162–166

---

3. ∠NMO is _____ (acute or obtuse).

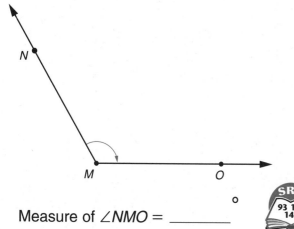

Measure of ∠NMO = _____°

SRB
**93 142
143**

---

4. Cross out the names that do not belong in the name-collection box below.

48
(2 * 3) * 8
100 − 62
18 + 13 + 17
12 * 4
184 ÷ 4

SRB
149

---

5. Round 451,062 to the nearest thousand. Circle the best answer.

A. 500,000

B. 451,000

C. 451,100

D. 452,000

SRB
182 183

---

6. Fill in the missing fractions on the number line.

SRB
316

LESSON 6·11

Math Boxes

---

1. Fill in the missing fractions on the number lines.

   a.
   $1\frac{2}{5}$ _____ $2\frac{2}{5}$

   b.
   $2\frac{1}{2}$ _____ 5

---

2. Draw 12 balloons. Circle $\frac{5}{12}$ of the balloons. Mark X on $\frac{1}{4}$ of the balloons.

   SRB 44

3. Write five names for $\frac{1}{4}$.

$\frac{1}{4}$

   SRB 149

---

4. a. Shade $\frac{5}{6}$ of the hexagon.

   b. Shade $\frac{2}{3}$ of the hexagon.

   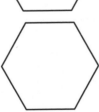

   SRB 44

5. Design a spinner such that it is more likely that you will land on red than on green.

   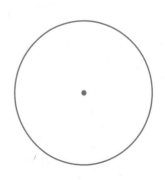

   SRB 84

My Route Log

Date	Country	Capital	Air distance from last capital	Total distance traveled so far
	1 U.S.A.	Washington, D.C.		
	2 Egypt	Cairo		
	3			
	4			
	5			
	6			
	7			
	8			
	9			
	10			
	11			
	12			
	13			
	14			
	15			
	16			
	17			
	18			
	19			
	20			

Route Map

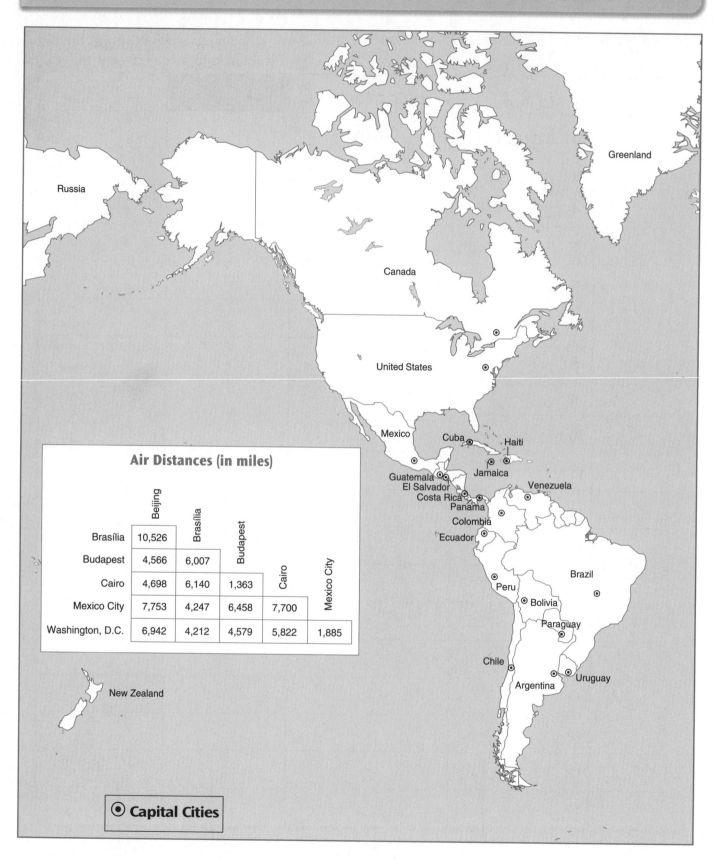

Russia

Greenland

Canada

United States

Mexico

Cuba

Haiti

Guatemala

Jamaica

El Salvador

Venezuela

Costa Rica

Panama

Colombia

Ecuador

Peru

Brazil

Bolivia

Paraguay

Chile

Uruguay

Argentina

New Zealand

Air Distances (in miles)

	Beijing	Brasília	Budapest	Cairo	Mexico City
Brasília	10,526				
Budapest	4,566	6,007			
Cairo	4,698	6,140	1,363		
Mexico City	7,753	4,247	6,458	7,700	
Washington, D.C.	6,942	4,212	4,579	5,822	1,885

◉ **Capital Cities**

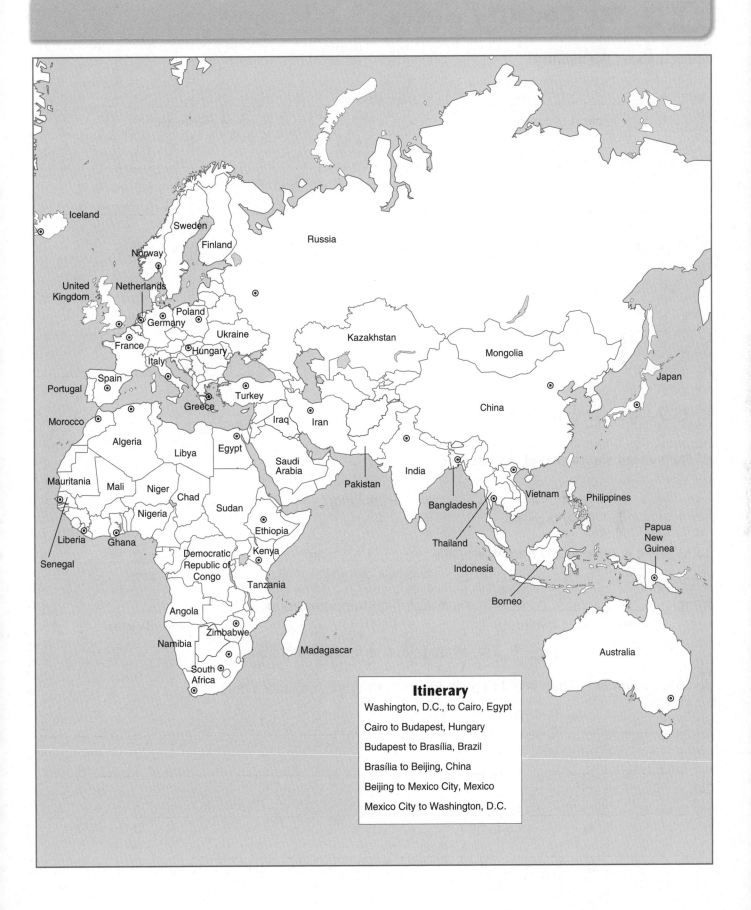

Itinerary

Washington, D.C., to Cairo, Egypt

Cairo to Budapest, Hungary

Budapest to Brasília, Brazil

Brasília to Beijing, China

Beijing to Mexico City, Mexico

Mexico City to Washington, D.C.

LESSON 3·6

My Country Notes

A. Facts about the country

_____ is located in _____.
name of country   name of continent

1. It is bordered by _____
   countries, bodies of water

   _____.

2. Population: _____   Area: _____ square miles

3. Languages spoken: _____

   _____

4. Monetary unit: _____

5. Exchange rate (optional): 1 _____ = _____

B. Facts about the capital of the country

_____   Population: _____
name of capital

1. When it is noon in my hometown, it is _____ in _____.
   time (A.M. or P.M.?)   name of capital

2. In _____/_____, the average high temperature in _____
   month   month   name of capital

   is about _____°F. The average low temperature is about _____°F.

3. What kinds of clothes should I pack for my visit to this capital? Why?

   _____

   _____

   _____

   _____

My Country Notes *continued*

4. Turn to the Route Map found on journal pages 172 and 173.
Draw a line from the last city you visited to the capital of this country.

5. If your class is using the Route Log, record the information on journal page 171 or *Math Masters,* page 421.

6. Can you find any facts on pages 302–305 in your *Student Reference Book* that apply to this country? For example, is one of the 10 tallest mountains in the world located in this country? List all the facts you can find.

_____

_____

_____

_____

_____

c. My impressions about the country

Do you know anyone who has visited or lived in this country? If so, ask that person for an interview. Read about the country's customs and about interesting places to visit there. Use encyclopedias, travel books, the travel section of a newspaper, or library books. Try to get brochures from a travel agent. Then describe below some interesting things you have learned about this country.

_____

_____

_____

_____

_____

_____

_____

LESSON 4·7

My Country Notes

A. Facts about the country

_____ is located in _____ .
name of country   name of continent

1. It is bordered by _____
   countries, bodies of water

   _____ .

2. Population: _____   Area: _____ square miles

3. Languages spoken: _____

   _____

4. Monetary unit: _____

5. Exchange rate (optional): 1 _____ = _____

B. Facts about the capital of the country

_____   Population: _____
name of capital

1. When it is noon in my hometown, it is _____ in _____ .
   time (A.M. or P.M.?)   name of capital

2. In _____/_____ , the average high temperature in _____
   month   month   name of capital

   is about _____°F. The average low temperature is about _____°F.

3. What kinds of clothes should I pack for my visit to this capital? Why?

   _____

   _____

   _____

   _____

176

**LESSON
4·7** **My Country Notes** *continued*

4. Turn to the Route Map found on journal pages 172 and 173.
   Draw a line from the last city you visited to the capital of this country.

5. If your class is using the Route Log, record the information on journal page 171 or
   Math Masters, page 421.

6. Can you find any facts on pages 302–305 in your *Student Reference Book* that
   apply to this country? For example, is one of the 10 tallest mountains in the world
   located in this country? List all the facts you can find.

_____

_____

_____

_____

_____

c. My impressions about the country

Do you know anyone who has visited or lived in this country? If so, ask that person
for an interview. Read about the country's customs and about interesting places to
visit there. Use encyclopedias, travel books, the travel section of a newspaper, or
library books. Try to get brochures from a travel agent. Then describe below some
interesting things you have learned about this country.

_____

_____

_____

_____

_____

_____

_____

LESSON 5·11

My Country Notes

A. Facts about the country

_____ is located in _____.

name of country   name of continent

1. It is bordered by _____

countries, bodies of water

_____.

2. Population: _____ Area: _____ square miles

3. Languages spoken: _____

_____

4. Monetary unit: _____

5. Exchange rate (optional): 1 _____ = _____

B. Facts about the capital of the country

_____ Population: _____

name of capital

1. When it is noon in my hometown, it is _____ in _____.

time (A.M. or P.M.?)   name of capital

2. In _____/_____, the average high temperature in _____

month   month   name of capital

is about _____°F. The average low temperature is about _____°F.

3. What kinds of clothes should I pack for my visit to this capital? Why?

_____

_____

_____

_____

LESSON 5·11 My Country Notes *continued*

4. Turn to the Route Map found on journal pages 172 and 173.
   Draw a line from the last city you visited to the capital of this country.

5. If your class is using the Route Log, record the information on journal page 171 or
   Math Masters, page 421.

6. Can you find any facts on pages 302–305 in your *Student Reference Book* that
   apply to this country? For example, is one of the 10 tallest mountains in the world
   located in this country? List all the facts you can find.

_____

_____

_____

_____

_____

c. My impressions about the country

Do you know anyone who has visited or lived in this country? If so, ask that person
for an interview. Read about the country's customs and about interesting places to
visit there. Use encyclopedias, travel books, the travel section of a newspaper, or
library books. Try to get brochures from a travel agent. Then describe below some
interesting things you have learned about this country.

_____

_____

_____

_____

_____

_____

My Country Notes

A. Facts about the country

_____ is located in _____.
name of country   name of continent

1. It is bordered by _____
countries, bodies of water

_____.

2. Population: _____   Area: _____ square miles

3. Languages spoken: _____

_____

4. Monetary unit: _____

5. Exchange rate (optional): 1 _____ = _____

B. Facts about the capital of the country

_____   Population: _____
name of capital

1. When it is noon in my hometown, it is _____ in _____.
time (A.M. or P.M.?)   name of capital

2. In _____/_____, the average high temperature in _____
month   month   name of capital

is about _____°F. The average low temperature is about _____°F.

3. What kinds of clothes should I pack for my visit to this capital? Why?

_____

_____

_____

_____

My Country Notes *continued*

4. Turn to the Route Map found on journal pages 172 and 173.
   Draw a line from the last city you visited to the capital of this country.

5. If your class is using the Route Log, record the information on journal page 171 or
   Math Masters, page 421.

6. Can you find any facts on pages 302–305 in your *Student Reference Book* that
   apply to this country? For example, is one of the 10 tallest mountains in the world
   located in this country? List all the facts you can find.

_____

_____

_____

_____

_____

c. My impressions about the country

Do you know anyone who has visited or lived in this country? If so, ask that person
for an interview. Read about the country's customs and about interesting places to
visit there. Use encyclopedias, travel books, the travel section of a newspaper, or
library books. Try to get brochures from a travel agent. Then describe below some
interesting things you have learned about this country.

_____

_____

_____

_____

_____

_____

PROJECT 1

U.S. Traditional Addition 1

Algorithm Project 1

Use any strategy to solve the problem.

1. There are 279 boys and 347 girls at a school assembly. How many students are at the assembly?

   _____ students

Use U.S. traditional addition to solve each problem.

2.   559
    +   72

3.   3,743
    +   5,106

4.   328
    + 474

5. 1,885 + 6,167 = _____

6. _____ = 456 + 198 + 618

7. 5,506 + 4,677 = _____

U.S. Traditional Addition 2

Algorithm Project 1

Use U.S. traditional addition to solve each problem.

1. From Monday through Friday, Peng read
   388 pages of a book. On Saturday and
   Sunday, he read 159 more pages. How many
   pages did Peng read during the week?

   _____ pages

2.
```
   633
    92
 +  48
```

3.
```
    905
 +  496
```

4.
```
  2,553
+ 6,424
```

5. $5{,}714 + 5{,}789 =$ _____

---

6. _____ $= 4{,}343 + 526$

7. $3{,}766 + 9{,}469 =$ _____

PROJECT 1 # U.S. Traditional Addition 3

Algorithm Project 1

Use U.S. traditional addition to solve each problem.

1. Hiroshi had $356 in his bank account this morning. This afternoon he deposited $85 into the account. How much is in Hiroshi's account now?

   $ _____

2. Write a number story for 448 + 375.
   Solve your number story.

   _____

   _____

   _____

   _____

Fill in the missing digits in the addition problems.

3.
```
  □ □ □
  5 6 9 6
+ 3 6 7 8
─────────
□ 3 □ 4
```

4.
```
      □
  6 3 5 2
+ □ 4 9 □
─────────
7 8 □ 5
```

5.
```
  □ □
  4 9 4
+ 6 2 7
───────
1 □ 2 □
```

6.
```
1 □ 1
9 9 8 □
+ 1 4 9
─────────
1 □ 1 □ 5
```

PROJECT 1 — U.S. Traditional Addition 4

Algorithm Project 1

Use U.S. traditional addition to solve each problem.

1. Sara and James ran for school president.
In the election, 529 students voted for Sara,
and 378 voted for James. How many students
voted in the election?

_____ students

2. Write a number story for 483 + 577.
Solve your number story.

_____

_____

_____

_____

Fill in the missing digits in the addition problems.

3.

```
    1   1
    5   6   3
+   2   9  ☐
  ☐   ☐   2
```

4.

```
  1   1   1
  8   9  ☐   9
+    ☐   0   2
☐    1   0   1
```

5.

```
☐   ☐   ☐
2   8   5   8
+   7   4   4   7
1   ☐   3   ☐   ☐
```

6.

```
        ☐   1
    4   0   0   4
+   8   6   9  ☐
1   ☐   ☐   0   0
```

PROJECT 2 | **U.S. Traditional Addition: Decimals 1**

Algorithm Project 2

Use any strategy to solve the problem.

1. Angela spent $2.62 at the craft store. She spent $3.94 at the fabric store. How much money did Angela spend in all?

   $ _____

Use U.S. traditional addition to solve each problem.

2. 7.69 + 38.5 = _____

3. _____ = 6.48 + 29.6

4. $9.59 + $0.45 = $_____

5. $30.45 + $65.99 = $_____

6. 54.11 + 9.2 = _____

7. _____ = 2.88 + 83.09

PROJECT 2

U.S. Traditional Addition: Decimals 2

Algorithm Project 2

Use U.S. traditional addition to solve each problem.

1. José had $5.98 in his wallet. He found 75¢ under his bed. How much money does José have now?

   $ _____

---

2. 3.9 + 4.48 = _____

3. 0.8 + 9.94 = _____

---

4. _____ = 6.76 + 28.18

5. 1.09 + 24.58 = _____

---

6. _____ = 1.03 + 52.81

7. 3.8 + 77.92 = _____

PROJECT 2

U.S. Traditional Addition: Decimals 3

Algorithm Project 2

Use U.S. traditional addition to solve each problem.

1. There is a flower growing in Kayla's garden. It was
   22.48 centimeters tall. In three months, it grew
   8.6 centimeters. How tall is the flower now?

   _____ centimeters

2. Write a number story for $3.80 + $5.12.
   Solve your number story.

   _____

   _____

   _____

   _____

Fill in the missing digits in the addition problems.

3.
```
      1   1
      3 . 8   5
  +     6 . □   7
   ┌──┬──┬─┬──┐
   │  │  │.5│  │
   └──┴──┴─┴──┘
```

4.
```
   □    1   1
      4 9 . 0   9
  +       6 . □   □
   ┌──┬──┬─┬──┐
     5 □ . 0   0
```

5.
```
   □      1
     2 9 . □   2
  +      0 . 9 □
   ┌──┐
     □ 0 . 3   9
```

6.
```
       1   1
     7 □ . 4 □
  +  1 2 . 8   6
   ┌──┐   ┌──┐
   │  │ 4 .│  │ 0
```

PROJECT 2

U.S. Traditional Addition: Decimals 4

Algorithm Project 2

Use U.S. traditional addition to solve each problem.

1. Surina and Lee are saving their money. Surina has $18.63. Lee has $24.81. How much money do they have altogether?

   $ _____

2. Write a number story for 9.8 + 48.36.
   Solve your number story.

   _____

   _____

   _____

   _____

Fill in the missing digits in the addition problems.

3.
```
   ☐ ☐
   5 0 . 3 5
 +   9 . 7 0
 ─────────────
   ☐ 0 . 0 ☐
```

4.
```
         1
     9 . 1 8
 +   2 . ☐ ☐
 ─────────────
   ☐ ☐ . 9 1
```

5.
```
   1       1
   7 9 . 0 7
 +   4 4 . ☐ 5
 ─────────────
   ☐ 2 ☐ . 4 ☐
```

6.
```
   1 1 ☐
   2 5 . 3 2
 + 2 ☐ . 7 9
 ─────────────
   ☐ ☐ 0 . 1 ☐
```

PROJECT 3

U.S. Traditional Subtraction 1

Algorithm Project 3

Use any strategy to solve the problem.

1. A store has 625 shirts and 379 pairs of pants. How many more shirts does the store have?

   _____ shirts

Use U.S. traditional subtraction to solve each problem.

2.   325
    −   68

3.   613
    −   249

4.   1,544
    −   749

5. 3,651 − 1,995 = _____

6. _____ = 506 − 187

7. 7,003 − 4,885 = _____

PROJECT 3

U.S. Traditional Subtraction 2

Algorithm Project 3

Use U.S. traditional subtraction to solve each problem.

1. The drive to Yuri's grandmother's house is 642 miles. Yuri's family has driven 484 miles so far. How many miles do they have left to drive?

_____ miles

---

2.   860
    − 86

3.   707
    − 389

---

4.   595
    − 397

5. _____ = 6,113 − 876

---

6. _____ = 4,552 − 1,688

7. 8,207 − 3,579 = _____

PROJECT 3

U.S. Traditional Subtraction 3

Algorithm Project 3

Use U.S. traditional subtraction to solve each problem.

1. Althea has 233 bean-bag animals.
   79 of them are bears. How many of her
   bean-bag animals are not bears?

   _____ bean-bag animals

2. Write a number story for 505 − 267.
   Solve your number story.

   _____

   _____

   _____

   _____

Fill in the missing numbers in the subtraction problems.

3.

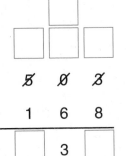

4.

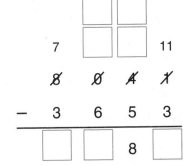

5.

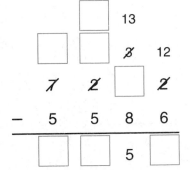

6.

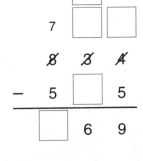

PROJECT 3

U.S. Traditional Subtraction 4

Algorithm Project 3

Use U.S. traditional subtraction to solve each problem.

1. Shane has $278 in his bank account. Caitlin
   has $425 in her bank account. How much
   more does Caitlin have in her account?

   $ _____

2. Write a number story for 503 − 347.
   Solve your number story.

   _____

   _____

   _____

   _____

Fill in the missing numbers in the subtraction problems.

3.
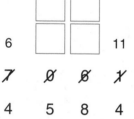

4.
```
       □
    □  □  □
    7  ∅  2
 −  2  6  8
 ─────────
    □  3  □
```

5.
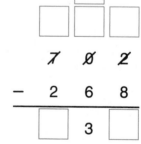

6.
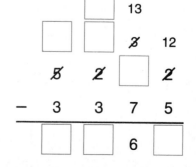

PROJECT 4

U.S. Traditional Subtraction: Decimals 1

Algorithm Project 4

Use any strategy to solve the problem.

1. Seth paid $6.72 for his lunch. Lily paid $3.79
   for her lunch. How much more did Seth's
   lunch cost?

   $ _____

Use U.S. traditional subtraction to solve each problem.

2. $9.75 - 4.32 =$ _____

3. $5.06 - 2.49 =$ _____

4. _____ $= 8.2 - 5.36$

5. $34.27 - $16.38 = $$ _____

6. _____ $= 50.08 - 27.39$

7. $6.35 - 2.37 =$ _____

PROJECT 4

U.S. Traditional Subtraction: Decimals 2

Algorithm Project 4

Use U.S. traditional subtraction to solve each problem.

1. Joanna had $73.48 in her bank account. She wrote a check for $25.69. How much money is in her bank account now?

$ _____

---

2. 6.04 − 2.75 = _____

3. 8.73 − 4.21 = _____

---

4. _____ = 5.63 − 2.64

5. 31.5 − 7.82 = _____

---

6. $_____ = $45.26 − $26.37

7. 60.08 − 43.29 = _____

Date _____ Time _____

Algorithm Project 4

Use U.S. traditional subtraction to solve each problem.

1. Riley bought two card games at the store. The total
   cost (before tax) was $9.25. One game cost $3.89.
   How much did the other game cost?

   $ _____

2. Write a number story for $38.42 − $19.76.
   Solve your number story.

   _____

   _____

   _____

   _____

Fill in the missing numbers in the subtraction problems.

3.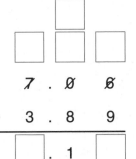

```
      ☐
   ☐  ☐  ☐
   7 . 0̸ 6̸
 − 3 . 8  9
   ☐ . 1  ☐
```

4.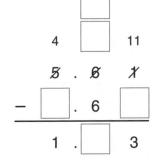

```
         ☐
    4  ☐   11
    5̸ . 6̸  7̸
 −  ☐ . 6  ☐
    1 . ☐  3
```

5.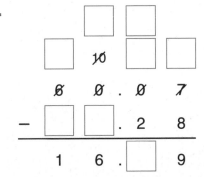

```
      ☐   ☐
   ☐  1̸0̸   ☐
   6̸  0̸ . 0̸  7
 − ☐  ☐ . 2  8
   1  6 . ☐  9
```

6.

```
       ☐    ☐
    3  4̸  ☐  10
    4̸  5̸ . 4̸  ☐
 −  ☐    8 . 8  5
    2  ☐ . ☐   5
```

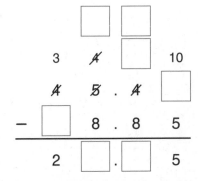

PROJECT 4

U.S. Traditional Subtraction: Decimals 4

Algorithm Project 4

Use U.S. traditional subtraction to solve each problem.

1. Quinn has two pieces of ribbon. The yellow ribbon is 12.42 meters long. The pink ribbon is 16.75 meters long. How much shorter is the yellow ribbon?

    _____ meters

2. Write a number story for 7.63 − 1.84. Solve your number story.

    _____

    _____

    _____

    _____

Fill in the missing numbers in the subtraction problems.

3.

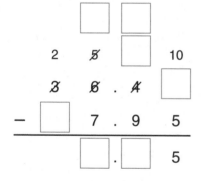

4.

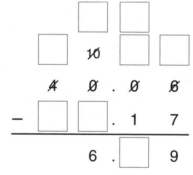

5.

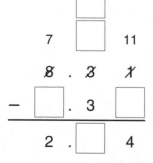

6.

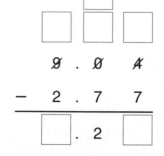

PROJECT 5

U.S. Traditional Multiplication 1

Algorithm Project 5

Use any strategy to solve the problem.

1. Mountain View Elementary School held a food drive.
   Each student donated 4 cans of food. There are
   676 students at the school. How many cans of food
   did the students donate altogether?

   _____ cans

Use U.S. traditional multiplication to solve each problem.

2. 2 * 413 = _____

3. 265 * 4 = _____

4. _____ = 46 * 307

5. 278 * 43 = _____

6. 18 * 72 = _____

7. _____ = 459 * 40

PROJECT 5 | # U.S. Traditional Multiplication 2 |

Algorithm Project 5

Use U.S. traditional multiplication to solve each problem.

1. The Riveras' cornfield has 75 rows. Each row
   contains 256 corn plants. How many corn plants
   do the Riveras have in all?

   _____ corn plants

2. $64 * 6 =$ _____

3. $213 * 30 =$ _____

4. $492 * 8 =$ _____

5. $70 * 572 =$ _____

6. $3 * 359 =$ _____

7. _____ $= 63 * 36$

PROJECT 5

U.S. Traditional Multiplication 3

Algorithm Project 5

Use U.S. traditional multiplication to solve each problem.

1. A machine can fill 258 bottles of juice per minute. How many bottles can the machine fill in 45 minutes?

   _____ bottles

2. Write a number story for 725 * 6.
   Solve your number story.

   _____

   _____

   _____

   _____

Fill in the missing digits in the multiplication problems.

3.
```
    [ ]   5
  4   2   9
*         6
─────────────
  2  [ ]  7  [ ]
```

4.
```
     3   2
     4   3
     3   6   5
*        5   7
──────────────────
     2   5  [ ]  5
+ 1 [ ] [ ]  5   0
──────────────────
  2   0  [ ]  0   5
```

5.
```
         [ ]
         2
         6   4
*            4  [ ]
──────────────────
     3   8   4
+ [ ] [ ]    6   0
──────────────────
  [ ]   9   4  [ ]
```

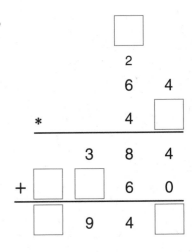

PROJECT 5 — U.S. Traditional Multiplication 4

Algorithm Project 5

Use U.S. traditional multiplication to solve each problem.

1. The zebra at the city zoo weighs 627 pounds. The hippopotamus weighs 5 times as much as the zebra. How much does the hippopotamus weigh?

   _____ pounds

2. Write a number story for 584 * 23.
   Solve your number story.

   _____

   _____

   _____

   _____

Fill in the missing digits in the multiplication problems.

3.
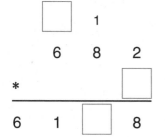

4.
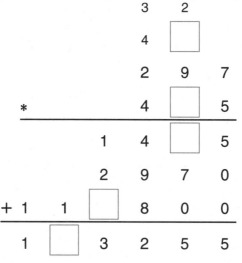

5.
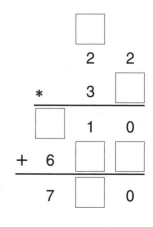

Algorithm Project 6

Use any strategy to solve the problem.

1. A turkey sandwich at Jason's Deli costs $5.98.
What is the cost of 4 turkey sandwiches?

$_____

Use U.S. traditional multiplication to solve each problem. Use estimation
or count decimal places to place the decimal point in your answers.

2. 12.64 * 5 = _____

3. $9.12 * 23 = $_____

4. $_____ = 86 * $0.57

5. 3 * $45.80 = $_____

6. _____ = 50.7 * 65

7. 426 * 5.3 = _____

PROJECT 6

U.S. Traditional Multiplication: Decimals 2

Algorithm Project 6

Use U.S. traditional multiplication to solve each problem. Use estimation or count decimal places to place the decimal point in your answers.

1. Find the area of the rectangle.

_____ m²

5 m ▭
      24.36 m

2. 18 * 30.09 = _____

3. $24.05 * 6 = $_____

4. _____ = 34 * 0.67

5. $8.53 * 76 = $_____

6. _____ = 2.3 * 5,084

7. $5.21 * 4 = $_____

U.S. Traditional Multiplication: Decimals 3

Algorithm Project 6

Use U.S. traditional multiplication to solve each problem. Use estimation or count decimal places to place the decimal point in your answers.

1. The average weight of a beagle puppy at birth is about 0.25 kg. At 6 months, a male beagle can weigh about 32 times as much. About how much can a 6-month-old male beagle weigh?

_____ kg

2. Write a number story for 4.6 ∗ 28.
   Solve your number story.

   _____

   _____

   _____

   _____

Fill in the missing digits in the multiplication problems.

3.

```
          ☐
          ☐
      8 . 6   2
  *       4   3
    2  ☐  8  ☐
+ ☐ 4  4  ☐  0
  3  ☐  0 . 6  ☐
```

4.
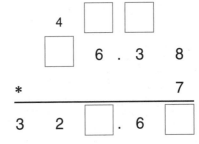

```
    4  ☐  ☐
  ☐    6 . 3   8
*                7
3  2  ☐ . 6  ☐
```

U.S. Traditional Multiplication: Decimals 4

Algorithm Project 6

Use U.S. traditional multiplication to solve each problem. Use estimation or count decimal places to place the decimal point in your answers.

1. Alicia has 7 pieces of yarn. Each piece is 3.65 meters long. What is the combined length of all 7 pieces?

_____ m

2. Write a number story for 5 * $48.30. Solve your number story.

_____

_____

_____

_____

Fill in the missing digits in the multiplication problems.

3.

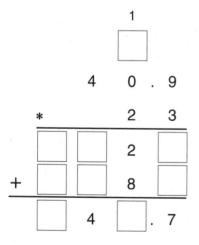

4.

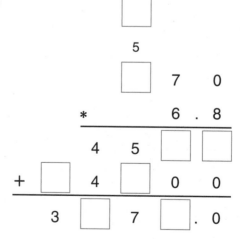

PROJECT 7

Long Division with One-Digit Divisors

Algorithm Project 7

Use any strategy to solve the problem.

1. The fourth-grade classes at Glendale School put on puppet shows for their families and friends. Ticket sales totaled $532, which the four classes are to share equally. How much should each class get?

   $ _____

   Be ready to explain how you found your answer.

Use U.S. traditional long division to solve each problem.

2. 78 / 6 = _____

3. 288 / 8 = _____

4. _____ = 564 / 3

5. _____ = 763 / 7

PROJECT 7

Long Division with One-Digit Divisors *cont.*

Algorithm Project 7

6. 350 / 4 → _____

7. 802 / 9 → _____

8. _____ ← 869 / 7

9. _____ ← 874 / 5

Go to www.everydaymathonline.com for additional practice pages.

PROJECT 7

Long Division with One-Digit Divisors *cont.*

Algorithm Project 7

10. Eight people visited a marine theme park. The total cost of the single-day admission tickets was $424. What was the cost per ticket?

$ _____

11. A national park charges an entrance fee of $3 per person. A school group visited the site. The cost was $288. How many people were in the school group?

_____ people

12. A family went on a six-day boat cruise. They sailed a total of 432 miles. They sailed the same distance each day. How far did they travel each day?

_____ miles

13. Four friends have birthdays in the same month. They decide to rent a hall to have a birthday party and split the cost evenly. The cost of renting the hall for one day is $172. How much did each friend pay?

$ _____

PROJECT 8 — Long Division with Larger Dividends

Algorithm Project 8

Use any strategy to solve the problem.

1. Four friends were playing a board game. Jen had
   to leave to go to her piano lesson. The three other
   players decided to divide Jen's money equally.
   Jen had $4,353. How much should each of the
   three other players get?

   $ _____

   Be ready to explain how you got your answer.

Use U.S. traditional long division to solve each problem.

2. $5,385 / 5 = $_____

3. $7,896 / 6 = $_____

4. _____ = 8,575 / 7

5. _____ = 8,127 / 3

PROJECT 8

Long Division with Larger Dividends *cont.*

Algorithm Project 8

Fill in the missing numbers.

6.

```
         1 □ 3 9
     5 ) 8 6 9 5
       - 5
         3 6
       - 3 5
           □ 9
         - 1 5
             4 □
           - 4 5
               0
```

7.

```
         5 □ □
     6 ) 3 2 5 2
       - □ 0
           2 □
         - 2 4
             1 2
           - 1 □
               0
```

8. Jai is saving money to go to sleep-away camp next summer. The total cost is $1,092. He is earning money by walking dogs in his neighborhood.

   a. At $4 per walk, how many dogs will Jai need to walk to earn $1,092?

   _____ dogs

   b. At $7 per walk, how many dogs will Jai need to walk to earn $1,092?

   _____ dogs

PROJECT 8 **Long Division with Dollars and Cents**

Algorithm Project 8

1. Dennis solved $9.45 / 7 like this.

   a. Study Dennis's work.

   b. Explain to your partner how he solved the problem.

```
        1.3 5
     7)9.4 5
      -7
       ‾‾
       2 4
      -2 1
       ‾‾‾
         3 5
        -3 5
         ‾‾‾
           0
```

Solve these division problems using Dennis's method.

2. $8.92 / 4 = $_____

3. $7.56 / 6 = $_____

4. _____ = 15.76 / 8

5. _____ = 19.17 / 9

*,/ **Fact Triangles 1**

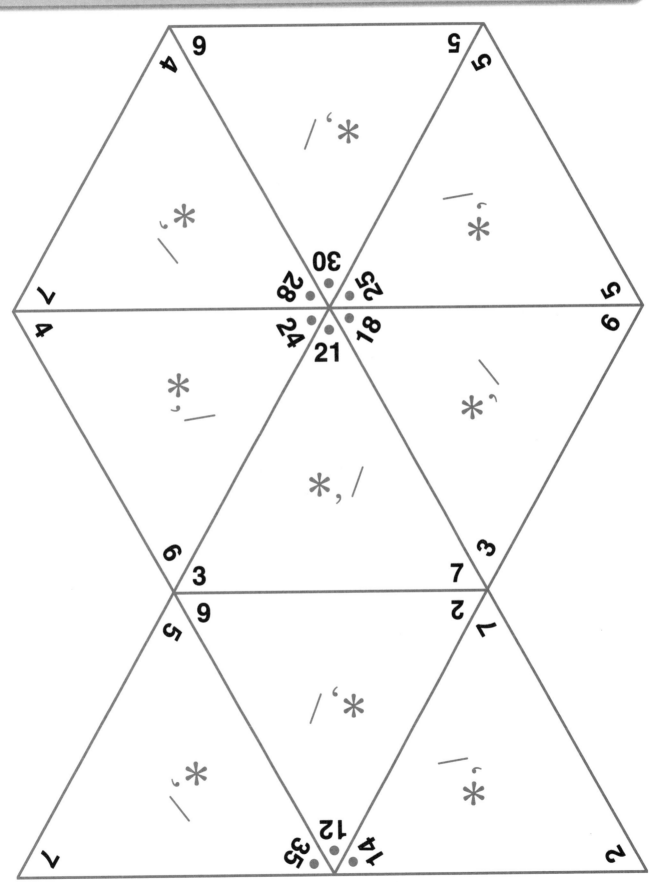

*,/ Fact Triangles 2

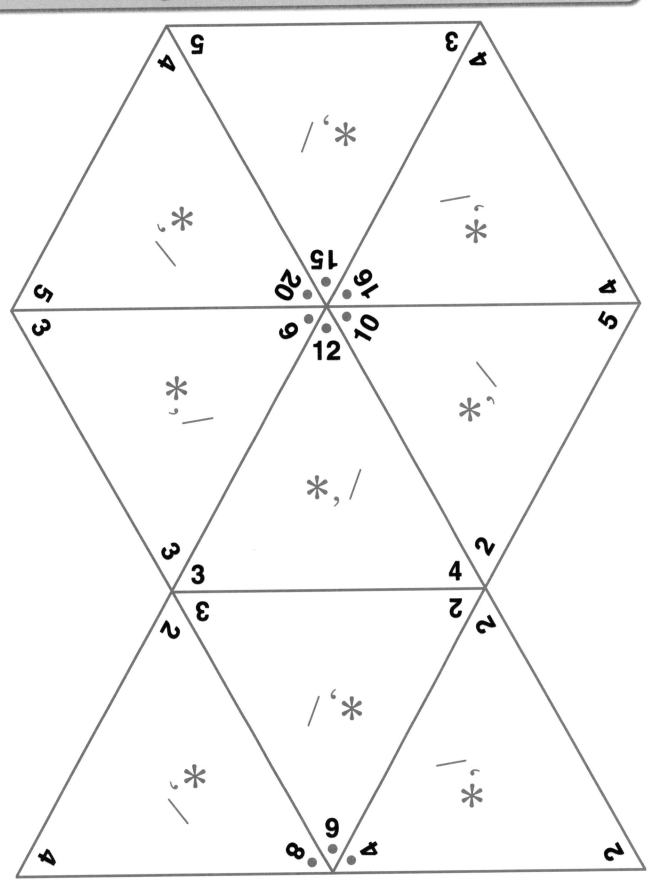

∗,/ **Fact Triangles 3**

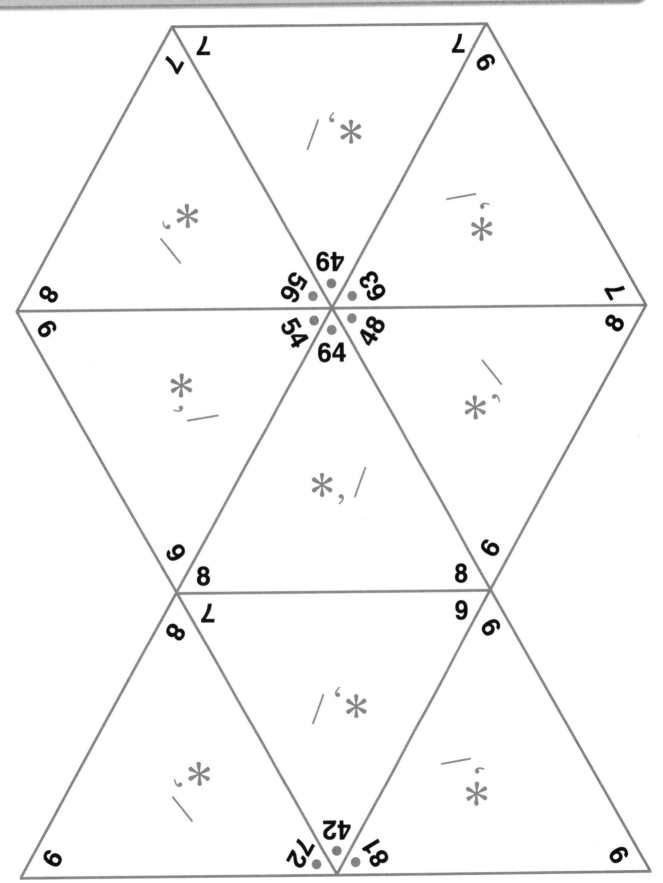

*,/ **Fact Triangles 4**

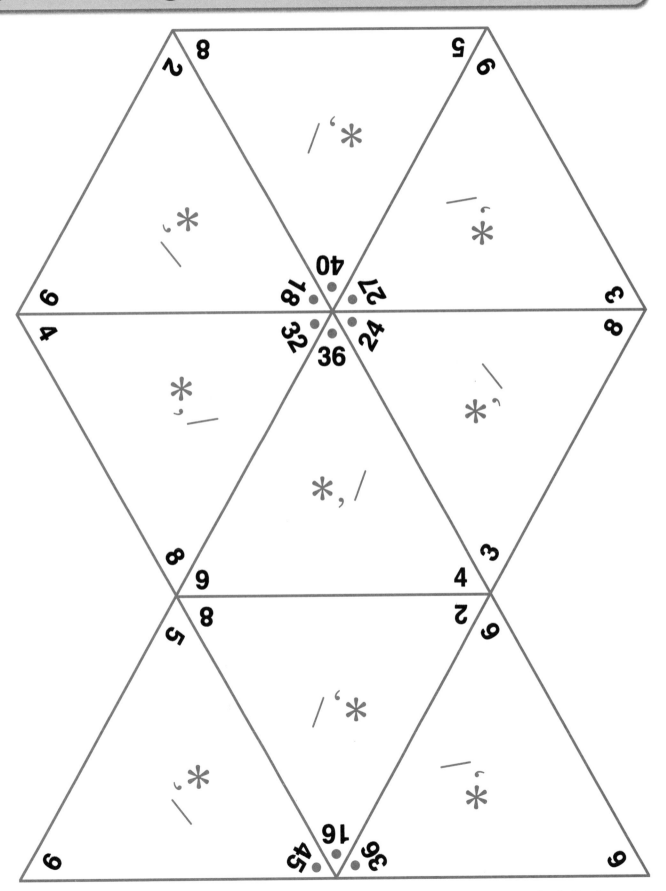